Anglo-Saxon Literary Landscapes

Anglo-Saxon Literary Landscapes

Ecotheory and the Environmental Imagination

Heide Estes

Routledge
Taylor & Francis Group
LONDON AND NEW YORK

First published in 2017 by Amsterdam University Press Ltd.

Published 2025 by Routledge
4 Park Square, Milton Park, Abingdon, Oxon OX14 4RN
605 Third Avenue, New York, NY 10158

Routledge is an imprint of the Taylor & Francis Group, an informa business

ISBN: 9789089649447 (hbk)
ISBN: 9781041175582 (pbk)
ISBN: 9781003690856 (ebk)

NUR 617 | 684 | 940

Cover illustration: © Douglas Morse

Cover design: Coördesign, Leiden

DOI 10.5117/9789089649447

For Product Safety Concerns and Information please contact our EU representative:
GPSR@taylorandfrancis.com
Taylor & Francis Verlag GmbH, Kaufingerstraße 24, 80331 München, Germany

Table of Contents

Acknowledgments

My mother and father gave me a love for granite and birch trees and the view from above treeline, and have led by example in their efforts to limit development and maintain the habitat of the frogs and the mosquitoes and the white-tail deer. My debts to them are beyond measure.

I thank Sherry Xie, inter-library loan wizard at Monmouth University, who conjured numberless items from the ether with good humor and speed. My colleagues in the English Department at Monmouth University have given me the time and the space to write, and the faith that I could do so; I thank in particular Kristin Bluemel, Susan Goulding, and Sue Starke. My students in courses on literature and the environment gave me an opportunity to begin thinking through some of the ideas that have come to fruition in this book.

At Clare Hall, University of Cambridge, Alison Gill, Laurie Zoloth, Rosanna Cantavella, Trudi Tate, and Lisa Mitchell listened patiently as I worked through the ideas in this book. Richard Dance, Judy Quinn, Elizabeth Ashman Rowe and the rest of the faculty of the Department of Anglo-Saxon, Norse and Celtic at Cambridge welcomed me for a second sabbatical, and the enthusiasm of ASNC's graduate students for my work was appreciated. Ruth Sternglantz, Laurel Chehayl, Harriet Soper, and Robert Stanton read portions of the work in progress and gave valuable feedback. Amara Hand and Ariana Tepedino helped with bibliographic details. Friends of more than forty years cheered me on through the ether: Melanie Blake, Doug Clay, and Heather Hodgkins believed in me and in this project, even when I did not. Ilse Schweizer Van Donkelaar of Amsterdam University Press encouraged, cajoled, and waited patiently for the completion of the manuscript. Gillian Rudd, AUP's peer reviewer, gave generous and encouraging feedback that improved the book immensely. A book on environmental criticism has to include a nod to the family dog, Jojo, who slept on my lap or at my feet for hours and hours upon end.

Dedication

This book could not have come to be without Douglas Morse, who has helped clear space and time for me to write this book, and has provided encouragement and moral support as well as cooking and cleaning along the way. He and Zeke have given up many nights and weekends and have cheerfully tolerated my absent-minded professorhood so that this project could come to be. To them, I dedicate this book.

1 Introduction

Environmentalists today worry about a newly felt sense of impermanence around places in which we live, arguing that we live in archipelagic, disconnected dwelling places in a time of increasing travel, migration across and among continents, and the construction of mass-market 'non-spaces' (Buell 2005: 63) such as fast-food joints and airports, indistinguishable one from another. Ecologists insist on the importance of seeing the environment not as a static background for human actions but as a system in flux. Post-colonial theorists point to the problems with treating not only places but also humans themselves as 'resources' for the fulfillment of other people's desires.

These may seem modern responses to modern problems. But Old English poems already convey a sense of place as impermanent, threatened by natural forces, by human acts of war, and by acts of God. The colonizing seizure of land that is interpreted simultaneously as both unoccupied and as occupied only by demons coexists in the surviving corpus of Old English texts with animals and trees defying domination by human enemies. The description of landscapes as existing in processes of change anticipates modern environmental observations. Old English poetry can be described as archipelagic in its survivals: we usually know little if anything about authorship, about place of composition, or about date, so each surviving poem (occasionally a group of poems) forms a small island on which scholars construct paleographical, critical, theoretical edifices, with some distant connections to one another as well as to prose works in Old English or Latin.

Richard Kerridge poses a valuable question about what genres and forms of literature can lead to environmental engagement. Thinking from ecotheoretical points of view (the plural is intentional) while reading Anglo-Saxon poetry and prose, in Old English and Latin, can itself constitute environmental engagement, and can also encourage further action and activism. As Greg Garrard argues 'ecology and environmentalism are themselves the outcomes of specific institutional and political histories, which continue to inform, constrain, and deform both fields of endeavor today. It is necessary to historicize ecology, as well as learning from it' (Garrard 2014: 3). As Robert Watson notes, the ideas that enabled the Industrial Revolution, and that enable the continuing disregard of the environment, 'took shape hundreds of years ago and cannot be effectively addressed until they are understood' (Watson 40). Watson points to assumptions that the Romantic era was the starting point of problematic ideas about the environment, notes the same tendencies in scholarship of Renaissance literature, and acknowledges that what might be called 'environmental literature' is older even that that: 'from the earliest instances of epic, pastoral and georgic, literature has offered a critique as well as an expression of nostalgia for the inviolate natural world that has always been not quite with us' (40). Like Watson, Kerridge makes the point that ecocritical engagement is, for many scholars, itself a form of activism: ecocritics 'are searching for ways of getting people to care' (362).

Many ecocritics and environmental activists dismiss or ignore the medieval, or misrepresent it in discussions of the modern; I will not catalogue those instances. One important aspect of this project is to bring the medieval into dialogue with ecocriticisms, to see how this project can lead to new readings of old texts but also how old texts and old ideas can challenge ecocriticism to think more sharply about historical contexts and how they have led to the current crisis. In the introduction to *Why the Middle Ages Matter*, Chazelle et al. argue that people – scholars and others – can and in fact must learn specifically from the Middle Ages. The period is often dismissed or ignored, but it is a source of and an important point of transmission for many of our current social formulations and constructions. Although the volume does not include an essay on environmental issues, the editors point out in the introduction that waste, an indicator of production, fell off dramatically in the transition from the long-distance economy of the Roman empire to the more local economies of the early Middle Ages (12). The point has often been made that the United States and other countries with heavily mechanized and huge agricultural conglomerates need to return to localized agricultural and economic production. The example of the early Middle Ages is that a return to local economies is possible, perhaps even without catastrophic origins or consequences.

Old English poetry predates environmental criticism and nature writing by centuries and cannot be said to participate in the debates and dialogues about what constitutes nature writing and how environmentalists should read literary texts. Yet reading Old English texts with attention to environmental depictions and concerns allows for new readings and interpretations of those texts and also opens up the possibility of introducing more nuance into modern views on the environment.

Anglo-Saxon Landscapes: Archaeological and Historical Evidence

In this book, I undertake to investigate how the Anglo-Saxons conceived of their relationship to the land and its nonhuman creatures, as described in literary and documentary texts. In fact, no single such relationship is discernable across Old English and Anglo-Latin poetry and prose: a range of attitudes exists. Landscape is presented as a neutral background to human activities, as an ordered environment in which humans and other creatures live out their natural lives, as a brief and fairly grim way-station on the path to eternal bliss in heaven or eternal suffering in hell, as a contested space in which physical and spiritual battles take place, and as a hostile environment for human activities. In order to consider meaningfully the depictions of landscape in Old English and Anglo-Latin texts, it is essential to have an idea about what sort of landscape Anglo-Saxon authors and scribes actually lived in. The landscapes of England varied considerably across different regions and there is good evidence that the uses of landscape shifted and evolved during the Anglo-Saxon period, understood as ranging from the arrival of Germanic groups in the post-Roman period through the Norman Conquest and perhaps beyond.

That said, it is possible to make the very general observation that the Old English landscape was primarily agricultural, with fields used at different times to grow vegetables, graze animals, or cultivate grains. Margaret Gelling's place-name studies and the archaeological research of Della Hooke, Tom Williamson, Debbie Banham, and others shed light on the landscapes of Anglo-Saxon England, including agriculture, forest, and town. Much of England had been cleared for agricultural use before the arrival of the Anglo-Saxons, and little wilderness remained, though there were fenlands largely impenetrable to outsiders as well as relatively small areas of forest and, primarily in the north and west, territories too mountainous to cultivate efficiently.

According to archaeological and historical studies, under Roman occupation the landscape of Britain was farmed quite intensively because the influx of troops, administrators, and other Roman colonizers increased the population, resulting in the need for greater amounts of food. The Romans brought with them the resources for grain production, which requires capital investment in plows and beasts of burden as well as mills and the means to transport grain and flour. The initial clearing of the land requires a particularly high investment of labor, but plowing, planting, harvesting, and processing the grain necessitate on-going investment of human, animal, and economic resources.

In the post-Roman period, with the drop in population and decrease in capital, grain production decreased significantly, and it was once believed that wild forest grew back over large areas of the landscape (Rackham 7-11). More recent research, however, demonstrates instead that fields cleared by Neolithic and, later, Roman inhabitants were turned to use for grazing animals and remained clear of forest (Hooke 2010: 113). The population increased gradually during the Anglo-Saxon period and eventually rose enough to require increased production of cereal crops, and to provide the labor force and capital to enable it.

Old English laws and boundary charters make frequent reference to agricultural land. An often cited passage from the late seventh-century Laws of Ine, copied under King Alfred in the late ninth century, enumerates the penalties prescribed for burning or cutting trees, in particular when woodland was used to pasture pigs:

> Gif mon afelle on wuda wel monega treowa, & wyrð eft undierne, forgielde III treowu ælc mid XXX scillingum.... Gif mon þonne aceorfe an treow, þæt mæge XXX swina undergestandan, & wyrð undierne, geselle LX scillinga.
> If someone were to fell very many trees in a forest, and it afterward becomes clearly known, [he must] pay thirty shillings for every three trees.... Then, if someone cuts down one tree, under which thirty swine could subsist, and it becomes known, [he must] pay sixty shillings. (*Laws* 51)

Charters describing manorial, parish or other boundaries take, like legal texts, a utilitarian perspective on the characteristics of the terrain, providing enough information about topographical features, waterways and notable plant life such as hedges and large trees to identify the territory belonging to an individual, a monastery, or another social body. For example, a charter in which King Æthelred grants to Eynsham Abbey a large parcel of land lists landmarks including lakes, paths, stones, trees, and thorn-bushes (*Electronic Sawyer* S.9 911).

The *Anglo-Saxon Chronicle* takes a more complex view of landscape, including utilitarian and anthropocentric material much like that found in the laws and charters as well as passages that describe landscapes and natural phenomena out of what appears to be intrinsic interest. The annual composition of the entries creates an immediacy that precludes a long-term overview, and in many entries the snapshot effect found in the charters is echoed in the annual additions. The *Peterborough Chronicle* entry for 656, for instance, records the gift of lands to the minster in Peterborough, including 'þas landes 7 þas wateres 7 meres 7 fennes 7 weres 7 ealle þa landa þa þærabuton liggeð' ('the lands and the waters and meres and fens and seas and all the lands that lie thereabout,' *Two Saxon Chronicles* 30). Later, the administrators of the minster had apparently rented out some of these lands, as the entry for

852 lists wood, brush, faggots, ale, and bread to be provided to the monks in exchange for the land lease. Many *Chronicle* entries refer to cattle and crops, and record features of the landscape in terms of their value to humans. In addition, the *Anglo-Saxon Chronicle* describes natural phenomena such as lightning or comets, which it interprets in terms of human concerns, as ill omens of famine or attack from abroad. The entry for 975 records:

> On þam ilcan geare on herfeste æteowde cometa se steorra. & com þa on þam eaftran geare swiðe mycel hungor. & swyðe mænigfealde styrunga geond Angelcyn.
> In the same year in the fall the star comet appeared, and then in the next year came a very great hunger and very manifold disturbances throughout the English people. (*Two Saxon Chronicles* 121)

Other *Anglo-Saxon Chronicle* entries, however, present a view of nature that anticipates the environmental writing of Thoreau or Muir, and which Lawrence Buell might have found ecologically oriented as defined in his 1995 book, *The Environmental Imagination*, in which he described the kinds of literature he thought was fully engaged with environmental issues. He adduces four criteria, two of which are that 'the human interest is not understood to be the only legitimate interest' and that 'some sense of the environment as a process rather than a constant or a given is at least implicit in the text' (1995: 7, 8). The *Anglo-Saxon Chronicle* contains numerous references to natural phenomena separate from the mention of human endeavors, suggesting that they are of intrinsic interest. Moreover, the *Chronicle* describes changing environmental conditions at several points.

In many cases, such passages reference astronomical phenomena such as comets (678, 892, 905, 995) lunar and solar eclipses (744, 773, 806, 809, 904), and, perhaps, the aurora borealis (926, 979). For example, in 734 a chronicler reports without further comment: 'Her wæs se mona swilce he wære mid blode begoten' ('In this year the moon was as if it were covered with blood'). The chroniclers also note such earth-bound phenomena such as 'se myccla winter' ('the great winter,' A761), and a great wind (1053, *Two Saxon Chronicles* 44, 51, 182). Yet another entry notes laconically that 'wundorlice nædran wæron geseogene on Suðseaxna lande' ('wondrous snakes were seen in the land of Sussex,' *Two Saxon Chronicles* 51). There is no articulation of any relationship between the snakes and the human occupants of Sussex: no comment about their potential utility or danger, no sense that they betoken some other event. As with other reports of natural phenomena, the chronicler's attention to the snakes seems a consequence only of interest or curiosity in some aspect of the natural world for its own sake, not out of concern for its relevance or potential value or harm to humans. Descriptions of earthquakes and winds show, albeit implicitly, a natural environment in a state of flux, not static, and one that is of interest not simply as a setting for human activities but for its own sake. Environmentalists today recognize that the earth is a mutable organism, not a static setting for human movement; so, it appears, did the Anglo-Saxon chroniclers.

Other entries describe changes in the environment as a result of human activity or as resulting in problems for human inhabitants of the land, again demonstrating a sense of the natural world as changeable, and not simply an inert setting for human affairs. The *Peterborough Chronicle* entry for 936 records:

> Syððon com se biscop Aðelwold to þære mynstre þe wæs gehaten Medeshamstede, ðe hwilon wæs fordon fra heðene folce. ne fand þær nan þing buton ealde weallas & wilde wuda.
> Then the bishop Athelwold came to the monastery which was called Peterborough, which was earlier destroyed by heathen folk. He found nothing there but old walls and wild woods. (*Two Saxon Chronicles* 115)

Several entries written during the tenth century describe the destruction of agriculture and livestock by raiding Danish armies and note the subsequently occurring famine, though without making an explicit connection between the two. The entry for 1006 summarizes the effects of the army's repeated attacks: 'hi hæfdon ælce scire on WestSexum stiðe gemarcod mid bryne. & mid hergunge' ('they had bitterly marked every shire in Wessex with burning and with harrowing,' *Two Saxon Chronicles* 137). An entry recorded just before the Norman Conquest is even clearer in its recognition that human actions have had environmental consequences:

> & þa Ryðrenan men dydan mycelne hearme ... hi ofslogon men & bærndon hus. and corn. & namon eall þet orf þe hi mihton tocuman, þæt wæs feola þusend. & fela hund manna hi naman. & læddon norð mid heom. swa þæt seo scir. & þa oðra scira þæ ðær neah sindon wurdon fela wintra ðe wyrsan.
> And then northern men did great harm ... they killed men and burned houses and grain, and took all the cattle they could get, that was many thousands, and many hundreds of men they took and led north with them so that the shire and the other shires which were near there were made for many winters the worse. (*Two Saxon Chronicles* 193)

It is not entirely clear whether 'þe wyrsan' refers to the destruction of crops or the loss of cattle or men, but it is reasonable to interpret the passage as indicating the combined effects of all three as implicated in the change in the countryside. Whoever wrote this *Chronicle* entry observed and recorded the fact that war was bad for the environment, because human actions resulted directly in the destruction of dwellings, landscape, and animals.

The use of trees and thorn-bushes alongside stones and streets in boundary charters suggests that their Anglo-Saxon creators had a view of nature, and even of individual trees, as quite static. For a boundary charter to function, the landmarks it identifies need to stay in place. It must be said, however, that the people who wrote such charters had to work within the constraints of the terrain they were surveying, and had to use the details of the terrain to create the best possible record of the transaction. This is modified by the more nuanced view of the natural world that the *Anglo-Saxon Chronicle* presents, in which some features of the natural world are of intrinsic interest apart from human concerns, and in which landscape features are observed to change as a result of human intervention, generally warfare.

Environmentalists and, subsequently, ecocritics have taken considerable interest in pastoral landscapes as well as in the wilderness. But in Anglo-Saxon texts, there is little evidence of wilderness terrain. Arguing from absence is always dangerous, and this may reflect the simple fact that documentary texts concern themselves primarily with inhabited areas, but archaeological evidence also establishes the relative paucity of wild regions in England during the period. There is a single reference to 'wilderness' in the *Anglo-Saxon Chronicle*, in the entry for 60 BC: When the Romans entered the British Isles, 'Ða flugon þa Brytwalas to þam wudu fæstenum' ('Then the Britons fled to the wooded wasteland,' *Two Saxon Chronicles* 5). This wood may have been the forest of Andred, which became a refuge for the natives again in 477, when the Angles and Saxons came, and 'þær ofslogon manige Walas & sume on fleame bedrifon on þone wudu þe is nemned Andredes lege' ('there they slew many Welsh and drove some in flight into the wood that is called the forest of Andred,' *Two Saxon Chronicles* 5). By the year 1000, the forest of Andred was being used as pastureland, and its boundaries no longer functioned as a barrier to outsiders (Hooke 1998: 143, 145). Additional references to the wilderness appear in Felix's *Vita Guthlaci* as well as in the Old English versions of the *Life of Guthlac*, a saint who retired to a hermitage in the fenlands of East Anglia, probably in 699 (Colgrave 5). The wilderness landscape in *Beowulf* is located in a probably imagined rather than remembered Denmark, not actually in England. But these descriptions of the fens as a 'wilderness' reflect a cultural construction of the area as seen by people who lived elsewhere. Fenlands, impenetrable to outsiders, may look like wildernesses, but they were exploited for fishing and fowling, salt production, fuel from peat, and for pasturing animals during the growing season, when arable lands needed to be kept free of grazing animals (Hooke 1998: 170, 178-79).

The rural landscapes described in contemporary documentary texts, with agricultural production and the breeding of cattle, pigs, and other food-producing animals, are the actual landscapes of Anglo-Saxon England. Anglo-Saxon chroniclers and scribes would have lived in towns, monasteries, or rural environments, and not in the wilderness. Archaeological research likewise investigates areas in which human activity has occurred in Britain's past. The paucity of references to the wilderness in Old English documentary texts and archaeological evidence does not reflect a lack of study or records of the wilderness, but is based on the reality that there was simply not very much wilderness in Anglo-Saxon England.

Wilderness, then, is found in Old English texts almost exclusively in literary sources rather than in historical documents. Given the absence of wilderness in Anglo-Saxon England, it might seem surprising that Old English texts contain as many references to wilderness as they do. There are a very few references to desert in Old English prose, all involving locations outside of Anglo-Saxon England: the *Old English Orosius*, *The Letter of Alexander to Aristotle*, and the *Wonders of the East* all refer to wasteland and wild areas in Africa and Asia, but not in Europe. The majority of the references to wilderness in Old English literature occur in poetic texts based upon biblical events or saints' lives that take place in locations temporally and geographically far from Anglo-Saxon England.

Neolithic residents of the British Isles began clearing the country's primeval forests for agricultural use, and from the time of the Romans perhaps fifteen percent of the English landscape remained forested (Hooke 1998: 151). Moreover, the enclosed and therefore relatively remote forests devoted to royal hunting so well-known from later medieval texts were a rarity in Anglo-Saxon England, where instead forests were used at least seasonally for animal pasturage and were also managed fairly intensively as a source of timber through coppicing and pollarding, practices in which trees are cut back to a low or high stump during periods of dormant growth and then allowed to grow back for several years to be harvested again. Depending on the species, branches were cut in fall or winter to allow for subsequent regrowth; in either case, this allowed for the harvesting of wood for building and burning without killing the tree. Coppiced trees are cut near the ground, which means they then need to be protected from animals during the period of regrowth. Pollarding, on the other hand, cuts branches back to a high stump so that pigs, cattle and forage can graze underneath the trees, eating nuts in season and nibbling on low branches as well as undergrowth, allowing the higher parts of the trees to continue growing (Hooke 2010: 139-40). During the harsher winter, animals might be brought into a barn or enclosed field nearer to dwellings; arable fields might be protected for agricultural use during the summer growing season by driving animals to pasture in woodland.

By the end of the tenth century one can imagine a landscape with large areas of open fields, some farmed for vegetables and fruits and some for grains, with other areas used for animal grazing. Extensively managed forests provided timber for building homes and churches as well as fences and various utensils and items of furniture. Fires for cooking and heat would have been stoked with peat or soft coal as well as with various kinds of scrap and garbage, including broken wooden implements. Small towns and monasteries tended to have structures clustered closely together amidst fields and stands of wood. A few settlements were large enough to consider urban, but even London had a population estimated at a mere 10,000 at its highest during the Anglo-Saxon period.

Defining Ecocritical Terms

The foregoing discussion of the Anglo-Saxon natural landscape assumes that anyone knows what 'nature' means, or 'wilderness,' or 'environment,' or 'animal,' or even 'human.' Defining the terms used in an ecocritical discussion of Old English texts is complicated by the fact that many of these words are attested only in later forms of the language. 'Wilderness' sounds like an Old English compound, but while there are instances of 'wild' and 'deor' ('wild' and 'animal'), the compound formed of the two with the suffix 'nesse' is not actually recorded. Old English writers use the term 'weste' ('wilderness' or 'wasteland'), most frequently to reference the desert territories described in biblical texts and early saints' lives, translated into Old English from drastically different literary and environmental contexts. Lives of the English saints Cuthbert and Guthlac also use 'weste' to delineate the watery but withdrawn terrains where they located their hermitages: in one case, an island, in the other, a raised area bounded by marshy fenland.

The word 'nature' is first attested in English in about 1400 to mean the material world in opposition to humans (*OED, s.v.*). The Old English word 'cynd' (or 'gecynd') passed into Middle English as 'kynde' with a consistent meaning referring not to what modern people think of as the natural (non-human?) world, but to different classes of things or animals, and the qualities that belong to them. Old English texts use 'gecyndelic' to refer to the group of characteristics that belongs to a particular 'kind' of being or thing. Interestingly, 'ungecyndelic' means 'supernatural' and 'monstrous' (Bosworth-Toller, *s.v.*) as well as 'unnatural,' or not belonging naturally to a particular kind or class of beings. The meaning of 'cyndelic' in Old English is itself varied and perhaps slippery, even without resorting to the later borrowing, 'nature,' as a translation.

'Nature' in popular usage today refers to flowers, trees, animals, storms, and mountains, but not to humans, human buildings, food, clothing, computers, or books. But humans are, of course, also part of the natural world, evolutionarily continuous with other animals and dependent upon chlorophyll, bacteria, seeds, and bees for our very survival. Defining the term 'nature' in modern theoretical work turns out to be as complicated as figuring out what qualifies as an Old English equivalent. Huggan and Tiffin comment that the difficulty in defining the term 'is compounded by the widespread perception that modernity, however defined, is 'post-natural' in the dialectical sense of losing human connection to the natural environment while simultaneously gaining a reinvigorated awareness that nature itself is continually reformed' (203).

Kylie Crane distinguishes 'nature' from both 'environment' and 'landscape,' defining the latter as 'a deeply cultural product' associated with 'the specifically visual or a tradition as manifested in visual arts' (10). She contrasts this with 'environment,' which she uses 'to designate all perceivable aspects of the physical world that surrounds a perceiving entity' (10). Environment, then, includes both natural and built terrain, but requires a human being (or, perhaps, animal or artificial intelligence) at its center. Landscape also assumes a (human) viewer, but is separate from rather than surrounding and encompassing the beholder (9). Crane uses 'nature,' in contrast, to mean something 'deliberately vague,' but distinct from and beyond what is understood under 'landscape' or 'environment' (12).

Clearly, 'nature' remains difficult to define today. Many ecocritics refer to 'non-human' nature, thus acknowledging that humans are part of nature while bracketing off all that is human as distinct in some way. As Gillian Rudd notes, even 'green' is problematic: while people today think of it as the color of nature and of environmentalism, in the late Middle Ages 'green' carried the connotation of inconstancy, in contrast to blue, representative of fidelity. 'For Chaucer and his contemporaries... green was the color of falsehood, unreliability, and deception, as well as the color of the natural world and of vigorous new life' (Rudd 2014: 30). Such a meaning is not attested for the Anglo-Saxon period, when 'grene' is used to refer to the color of grass and foliage as well as of gemstones and oxidized copper, as well as to vigorous (new) life and to unripe fruits or plants (*Dictionary of Old English*, *s.v.*). But the mutability of connotations and associations of 'green' even across the past six hundred years cautions against assuming continuity in the meanings of words.

Environmental Criticisms and Ecological Theories

'Ecocriticism' is a relatively new discipline within the humanities that investigates literary, historical, artistic, and other cultural depictions of the relationship(s) between humans and everything else. In its early evolution, 'ecocriticism' referred primarily to the literary depictions of natural environments and animals, but in more recent figurations it has migrated to disciplines dealing with material objects as well as documentary texts, and encompasses topics as diverse as cities and cyborgs, postcolonial theory and social justice. Ecocritics understand human activities as having caused harm to the earth and its non-human elements and creatures, and see the critical enterprise as engaged with efforts to reduce consumption and slow the processes of climate change. The extent to which ecocritics see their enterprise as explicitly political, or connect it with political activism, varies.

An important concept within environmental studies is the 'Anthropocene,' defined as the current geological age beginning when humans first impacted the environment. Scientists vary in where they locate the start of the Anthropocene, some arguing for the Industrial Revolution and the deposit of carbon in the earth's surface as a result of human activities, others for the nuclear age, when radioactive particles begin to appear in the earth as well as in human teeth; still others point to other watershed dates, as for instance the beginnings of the cultivation of plants or grains. Meanwhile, many ecocritics call the current era 'post-human,' referring to a time when humans can no longer ignore our effects on the planet and consider ourselves to be distinct from or exceptional in the context of the rest of the planet's creatures and things. As Eduardo Kohn argues in *How Forests Think*, 'Creating an analytical framework that can include humans as well as non-humans has been a central concern of science and technology studies' (6).

Trying to trace a history of environmental criticism at this point involves viewing with great excitement a field that is exploding in size even as it moves into critical and theoretical territories unexplored even five years ago, and seen by some readers (for example Dominic Head) as impossible. Because of the speed with which new environmental criticism is being produced in the humanities, any survey will be out of date by the time it is published. In the comments that follow, I identify some of the starting points of environmental engagements in the study of history, philosophy, and theology and some of the origins of environmental literary criticism, and then sketch some promising directions recent ecocriticism has taken. Importantly, 'environmental criticism' or 'ecocriticism' is not a singular approach to literature, art, and other cultural productions, but rather encompasses a wide variety of approaches and methods. Several scholars have anatomized points of overlap and tension among deep ecologists, ecofeminists, post-colonial ecocritics, and others. Serpil Oppermann comments:

> The only discernible pattern among ecocritical definitions is their focus on the importance of the relationship between literature and the physical environment; they also share the common aim to synthesize literary criticism with the natural sciences, and literary studies with the environmental philosophies. In fact as most of the ecocritics have repeatedly stated, ecocriticism seems to resist a single definition (105).

The biologist Rachel Carson's book about the effects of pesticides, *Silent Spring* (1962), is frequently cited as a point of origin for environmental criticism; certainly Carson brought the awareness of human effects on the environment to a large audience in a way that had not been done before. But she was not the first to observe environmental change or human effects on the environment. The postcolonial ecocritic Elizabeth Deloughrey argues that 'the rise of the modern concept of ecology and conservation... can also be attributed to the complex botanical networks of the eighteenth-century European colonial island laboratories' (323-34). Ecofeminists such as Val Plumwood (*Feminism and the Mastery of Nature*) and Carol Adams (*The Sexual Politics of Meat*) traced links between women's rights, animal rights, and environmental degradation; Peter Singer and PETA insisted on the sensibilities and therefore rights of animals, and Jacques Derrida discovered the gaze of his cat, as described in a series of lectures published as *The Animal That Therefore I Am*. A watershed moment for literary environmental studies was Cheryl Glotfelty and Harold Fromm's publication in 1995 of

The Ecocriticism Reader, which brought together a collection of essays that discussed ecological problems in European and American culture and looked to literary study as a source of engagement with environmental issues.

As a literary enterprise, ecocriticism began with two related but somewhat different approaches: re-reading canonical texts with a renewed focus on the presentation of the natural world and the place of the human in relation to it, and re-evaluating and bringing into the canon texts not previously read as literature. Eecocriticism explicitly identifies itself not merely as an aesthetics of reading but as a political movement concerned with environmental crisis and responses to it in literary works and by scholars of literature. These include a sense of a natural world in danger, overrun by industrialization and urban/suburban sprawl, and a concern with advocacy for the preservation of relatively undeveloped areas as 'wilderness.' In a 1999 *PMLA* Forum on environmental criticism, Patrick D. Murphy noted the move from earlier critical paradigms in which environments were understood simply as providing setting for the actions of characters: 'they are instead seen as a fundamental feature of the ideological horizons of literary works' (1099). Writing in the same issue of *PMLA*, Ursula Heise provided a working definition of ecocritical thinking: 'Ecocriticism analyzes the ways in which literature represents the human relation to nature at particular moments of history, what values are assigned to nature and why, and how perceptions of the natural shape literary tropes and genres. In turn, it examines how such literary figures contribute to shaping social and cultural attitudes toward the environment' (1097). Broadly speaking, ecocriticism investigates literary depictions of human engagements with the non-human world, as it both reflects and shapes cultures.

Early ecocriticism focused on literary texts that were fundamentally concerned with the 'natural' world: rural environments and wilderness areas. In his 1995 book *The Environmental Imagination: Thoreau, Nature Writing, and the Formation of American Culture* (1995), Lawrence Buell laid out several criteria for a work of literature to be deemed worthy of attention for environmental criticism. These include the ideas that nature must be a 'presence' in a work, not simply background or setting; that the work present an ethical sense of 'human accountability to the environment'; and that the environment be presented as an organism in a state of continual change, not a static entity (7-8). Buell's focus at this point was on literary works that engaged explicitly with issues of importance to late twentieth-century environmental activists. When the *Ecocriticism Reader: Landmarks in Literary Ecology* appeared a year after *The Environmental Imagination*, it included Buell's book in a list of recommended further reading and called it 'a monumental work' (394). The essays gathered in *The Ecocriticism Reader* focus, like Buell's book, on works in which nature is a profound presence, and include canonical texts by writers such as William Faulkner alongside those by 'nature writers' such as Edward Abbey, Annie Dillard and Henry David Thoreau, as well as the science-fiction novelist Ursula K. LeGuin.

Scholars of literature quickly broadened their scope, as attested by books with titles such as *Beyond Nature Writing* (ed. Armbruster and Wallace) and *The Nature of Cities* (ed. Bennett and Teague). Buell himself later recognized that for environmental criticism to move beyond a very small niche in the study of literature, it needed to expand its focus and consider how any literary work could be examined from an ecological point of view. In his 2005 book *The Future of Environmental Criticism*, Buell describes his own movement from analysis of nature writing to a concern with a broader field of literary genres and argues that 'a mature environmental aesthetics – or ethics, or politics – must take into account the interpenetration of metropolis and outback, of anthropocentric as well as biocentric concerns' (22-23). Important areas for analysis, in different times and places, are the relationships between humans and the environment, in whatever environments humans have built for themselves, and whether human interventions leave environments relatively unaltered or completely constructed. Rather than focusing upon pastoral and wilderness literature, or even upon representations of rural and wilderness landscapes in works not centrally concerned with nature, 'second-wave ecocriticism' (Buell 2005: 22) also examines human relationships with environments, animals, objects, and each other, in various built environments, including cities.

Ecocritical thinkers who write about the city open a window into the experiences of wilderness, rural and urban environments by varying groups of people. A problem with pastoral, twentieth-century American nature writing, and first-wave ecocriticism alike is that all are written from the point of view of privileged members of society: economically advantaged, generally male and Caucasian. 'Wilderness' in late twentieth-century environmental discourse is territory that is uninhabited by humans, or territory that from the point of view of environmental advocates *should* be uninhabited by humans, and should thus be preserved from development. The protection of wild regions or agricultural landscapes from strip-mining or the construction of suburban McMansions is hard to argue with. But in practice, rural inhabitants who may have worked the land in a particular region for generations or centuries have been dispossessed for the sake of preservation of land called 'wilderness' because its occupants are different in class and/or race from those doing the preservation. In the United States, for example, the creation of state and national parks has been done by removing the people who occupied the land before colonists arrived, as well as more recent settler farmers.

Writing about the city and about human entanglements with the natural world becomes a way of addressing several interconnected issues: the experiences of peoples of various economic classes and ethnic groups in and with wilderness and rural landscapes; the conception of wilderness in past generations as something essentially feminine – just a couple of examples might be the mountains in the United States Rockies named the Grand Tetons, or the giant breast-like mountain of H. Rider Haggard's *King Solomon's Mines* – viewed, possessed and controlled by male users; and the issue of environmental justice, which insists on the right of all people to have access to healthy living environments free of toxins and various kinds of pollution, whether in cities or in less densely populated areas.

Critics such as Michael Bennett, Karla Armbruster, Kathleen R. Wallace and Andrew Furman have explored the ways in which rural and wilderness landscapes have been irrelevant or even threatening for African Americans and Jews. In the ante-bellum United States, slaves living in towns and cities generally had better living conditions than those working in agriculture, since their living quarters were in owners' homes rather than in hovels among the fields and animals. Moreover, because slaves could not testify in court, the worst abuses of slavery in the United States occurred in areas where there were no white witnesses, i.e. away from towns and cities. The cities of the North became places to which slaves could escape from the plantations of the South and where they could, in effect, disappear, becoming anonymous among the large numbers of other humans, unlike in small towns where their appearance would surely be noticed. Jewish immigrants to the United States also settled in cities: the latest in a two-thousand-year history of migration to and expulsion from one nation after another. Forbidden from owning land in many medieval communities, Jews gathered in towns and cities where they could engage in trade, banking, and other occupations open to them. Furman suggests that the Jewish focus on texts, and indifference if not outright antipathy toward nature, can be traced even farther back, to traditional rabbinic wariness of pagan nature worship as well as to the creation of a Jewish community centered around the books of Torah following the destruction of the Temple in Jerusalem in the first century AD (52).

John Claborn points to the necessity of accounting for race in a cogent essay on W. E. B. Du Bois' travels to US national parks and 'wilderness' areas, described in 'Of Beauty and Death,' published in 1920. Claborn recounts Du Bois' description of the indignities of the Jim Crow car he was forced to ride in during his journey to the wilderness. 'By intertwining such seemingly disparate and opposed spaces Du Bois forces us to compare them according to the logic of double consciousness' (124). As Claborn notes 'Muir's writing also participates in a discourse that assumes a division between culture and nature – a division that Du Bois implicitly challenges as racially codified' (121). Claborn points out that this 'wilderness' was not naturally occurring, but established by removing the inhabitants from what had once been native territory, and was created as a commodity for white male explorers from the urban east and south. But Du Bois' description of the Grand Canyon sees a 'wounded, feminized earth.... Expressed as an act of phallic violence inflicted on the earth, the sublime functions as a strategy to recuperate and represent an authentic encounter with nature' (125-6). Claborn does not discuss the functions of gender in Du Bois' narrative, thus allowing an association between landscape and femininity to appear 'natural' rather than acknowledging that associations between the 'feminine' and earth or landscape are also cultural artifacts.

Ecofeminists investigate links between cultural constructions of the environment and of gender, and challenge such dualities in contemporary culture. The tendency to see humans as separate from the natural world, rather than as a part of it, is a culturally conditioned point of view. Val Plumwood argued that a set of dualisms originating at least as far back as ancient Greece links the feminine with nature in a hierarchy that privileges human over non-human, masculine over feminine, mind over body, and production over reproduction (72). This mentality conceives of women and the natural world as existing to satisfy human (male) needs and desires. Ecofeminists suggest, moreover, that for human beings to take a responsible position with regards to the natural world requires a shift from dialectic to dialogic thinking, a reconceptualization of ethical meanings from the perspective of relationship, continuity, and embrace of difference, rather than of paired oppositions (Murphy 1991 passim). As Stacy Alaimo argues, 'Feminist theory and gender studies have demonstrated... that many unmarked, ostensibly ungendered fields, modes, and sites of inquiry have been shaped by the social categories of gender, race, class, and colonialism' (Alaimo 2014: 188). Alaimo further argues that it is not possible to study science or environment without attending to 'the knots and entanglements that intertwine nature and culture, science and the humanities, the knower and the known' (Alaimo 2014: 188). But some ecofeminist studies have in turn been challenged by post-colonial ecocritics and other theorists of race and ethnicity as failing to attend to cultural, racial and economic differences in women's experiences, and thus focusing by default on the position of white women. Ecofeminism does not go far enough: it fails to recognize that it is not only women who become resources for men, but a wide variety of human beings, male and female, treated as 'resources' for those with privilege, who sometimes also include women.

Post-colonialist ecocritics bring together a wide variety of perspectives in showing how colonial ideologies about the exploitation of 'resources' have been problematic not only for the environment but also for many humans. As Deloughrey argues, 'Some of the work of postcolonial ecocriticism includes examining the implications of foundational narratives, problematizing assumptions of a universal subject and of an essentialized nature, and examining how forms of dominance are naturalized' (231). She points out that some environmentalists have taken the white male subject and its dominance as natural, and argues that colonization is not a subject for history alone, calling on ecocritics to attend to contemporary problems with United States military strategies, including the toxic environmental effects of various kinds of weapons and the tactics used in their deployment. She critiques Buell's claims that ecocriticism originated in the United States, and calls for '*contemporary* American ecocritics [to] use their strategic viewpoints to engage the ongoing military imperialism' (323). She notes the ways in which 'urban' and 'wilderness' have been intertwined since the beginnings of colonization, arguing that colonialism was 'a process that also occurred within and that radically changed the metropolitan center' (324). As discussed in Chapter 4, such colonization occurs in an Anglo-Saxon context centuries before modern 'exploration' began. As noted above, certain Anglo-Saxon texts show a recognition of the problem of environmental damage caused by military action. Deloughrey takes a broadly activist stance, arguing that ecocritics must attempt to 'dismantle the homogenizing networks of power in which we are enmeshed' (334).

Some work to connect environmental theory with disability and ableism has been done, notably in a conference on 'Composing Disability' held at George Washington University in April of 2016. Disability is interrelated with and sometimes directly caused by the design of constructed environments, while some disabling illnesses are caused directly by environmental degradation. As the conference organizers write 'Marginalized subjects, including disabled people, often experience their lives in greater proximity to environmental threats such as toxicity, climate change, generational exposures to unsafe living conditions due to poverty, militarization, [and] body exhausting labors'. This is an area where more work needs to be done.

Numerous environmentalists and ecotheorists have pointed to modern levels of material consumption as constitutive of environmental change, though few point to capitalism itself as the problem. John Bellamy Foster, however, argues that Karl Marx's writings pointed to a need to take environmental issues seriously. Moreover, Jonathan Maskit argues directly that environmental philosophers have not dealt with, or not attended sufficiently to, the problems of (over-)consumption. He observes, among environmentalists, two possible models for dealing with the problem that capitalism is designed to increase production and consumption without end: an individual model, following Arne Naess, which suggests 'that the knowledge that it is ecologically undesirable to consume more, or even as much as one does, will lead one simply to want to consume less,' and a political model, in which social policies should incentivize reduced consumption among individuals (Maskit 130). He argues, however, that neither model goes far enough:

> To say that there are downsides to modern life is surely not novel. What is new here is the suggestion that our addressing these concerns will require not merely technical, political, or policy suggestions, but a rethinking of what it means to be human. How could one at least begin to shape subjectivity? Here are some ideas: Don't watch television. Question all assertions that a practice is impossible. Know the people who produce your food. Figure out how to get from point A to point B without driving or flying. What is interesting about this list is that some of these things look like ascetic practices. And maybe they are. But they are practices oriented not towards being the way we always could have been but towards being a way that we did not know we could be (Maskit 140-41).

With its constant drive for MORE – more products, more resources, more consumers, more development – capitalism is deeply enmeshed with, if not directly causative of, environmental degradation: more production leads inevitably to more use of 'natural resources' and more waste. Reading and thinking from environmental perspectives are not enough: we need to take action at individual as well as communal levels.

The philosopher Timothy Morton (2013a) argues that it is important to understand climate change as 'hyperobject' which he defines as something so large that it cannot be apprehended by any individual at any one time, but only through the aggregation of large amounts of data across both time and space – which is why its existence can plausibly be denied. Scott Slovic argues that this is precisely why environmental study needs the humanities: stories provide a way to make sense of numbers too large for humans otherwise to comprehend. Cheryl Lousley calls for games that help players to understand human effects on the environment. Richard Kerridge contemplates what literary genres can best make environmental change real for readers. In *Prismatic Ecology* and *Stone*, Jeffrey Jerome Cohen moves beyond many of the usual concerns of ecocriticism. *Prismatic Ecology* argues that environmentalism focused on green misses far too much. The rainbow of essays that Cohen assembled engages with objects, queer theory, and ultraviolet light, among other subjects usually not pursued by ecocritics. In *Stone*, he intertwines scholarly engagement with personal writing in a way that is unusual in the field of literary study but might be crucial if scholars are to have any impact on human responses to climate change.

As noted above, any survey of contemporary ecocriticism and ecotheory is bound to be out of date before the book sees publication, because the field is expanding and developing so rapidly. I have attempted here to point to some key areas of engagement in a large and growing field rather than to provide a full survey: several books and volumes of essays have attempted to provide an overview of the field, but inevitably remain incomplete (see Garrard 2012 and 2014, Westling 2014, and Hiltner 2015, as well as earlier surveys edited by Branch and Slovic 2003 and Kerridge and Sammels 1998).

Two points, however, are crucial. First, scholarly ecocritical writers and writings should not maintain careful ivory-tower distance in an academic vacuum, but must engage with real-world environmental problems. Second, a responsible ecocriticism must be at the very least aware of, and ideally account explicitly for, the presence of a wide variety of humans on spectrums including ability, gender, class, religious affiliation, and race, and of how these human categories of difference intersect in ways important to our engagements with the non-human world.

Ecocriticism and Anglo-Saxon Studies

Ecocritical analyses of texts have included consideration of several broad environmental types, including representations of wilderness, animals, dwelling places and pollution, as well as pastoral and apocalyptic literary tropes. (See, for instance, Garrard 2012.) These major tropes, not surprisingly, do not map all that well onto Old English texts. Old English documentary texts make frequent references to rural landscapes in the form of discussions of arable land as well as the management (or plundering) of cattle, sheep, and other animals. There are few descriptions of earthly dwellings or of wilderness in documentary texts, though sermons discuss paradise and heaven, and poetry refers to Roman ruins as 'enta geweorc,' the work of giants, while wilderness appears regularly in Old English poetry. Old English texts make frequent and memorable reference to the sea, a major lived environment for Anglo-Saxons, who fished and sailed it, and were threatened by its deadly storms on land as well as at sea. But this is primarily a feature of poetry, including adaptations of Latin prose as well as Anglo-Saxon poems with no known antecedents, rather than of documentary texts.

Several scholars have made important studies of Anglo-Saxon literary environments that did not take an explicitly ecocritical perspective, in that they described landscapes and other aspects of the environment while treating depictions of 'nature' as separate from, and subordinate to, humans and human concerns. Jennifer Neville's 1999 monograph *Representations of the Natural World in Old English Poetry* presents a learned overview of representations of the environment in Anglo-Saxon poems in Latin and Irish contexts. Her focus, however, is on human existence within Old English textual environments, rather than on the environment as an independent entity with potential moral or ethical interests. Instead of being simply a stable backdrop for human activities, it is but is itself in flux. She sees nature as something which defines, confines, and constructs humans, both individually and in social contexts. Moreover, partly as a result of the assignment of the Grendel-kin of *Beowulf* to the natural world rather than something either human or monstrous, she finds the natural environment to be unremittingly hostile and threatening to humans.

In the 2006 volume *Literary Landscapes and the Idea of England, 700-1400*, Catherine Clarke devotes the first two chapters to an exploration of representations of nature in Old English literary texts. The works she considers most fully here are *Guthlac A*, with its lengthy representation of the wilderness into which Guthlac withdraws as a hermit, and Bede's description of the island of Britain in the prologue to his *Ecclesiastical History*, which she examines in the context of several other Old English texts as well as contemporary and earlier Latin texts. Clarke identifies the ways in which these literary works draw upon and disseminate classical notions of pastoralism, working within pastoral traditions to depict delightful landscapes. In contrast to *Guthlac A* and Bede's prologue, Clarke also examines several Old English poems in which pastoral conventions are employed in inverted form to create antagonistic literary landscapes; among these, she includes *The Wanderer* and *The Seafarer* as well as *Beowulf*. Her point here is to demonstrate fairly wide knowledge in Anglo-Saxon England of earlier pastoral conventions.

Clarke's presentation of the natural world is broader than Neville's, and her discussion of the potential delights of nature seems to align with celebrations of rural and wilderness landscape in the nature writing that early ecocriticism took as its primary subject. Her focus on the natural world, however, retains the sense that it provides an unchanging field for human activity, rather than being an organism, or set of interrelated organisms, that undergoes change or that is of interest in itself. In fact, an environmentally aligned analysis of literary texts finds the trope of pastoralism to be rather problematic. The literary form of pastoralism appears during a period of increasing urban development in classical Greece and idealizes the rural landscape as a location of peaceful escape opposed to the city, with intense human occupation of densely built-up spaces. In its classical beginnings, as in its later Renaissance and Romantic reappearances, pastoral literature tends toward a use of nature as a stage for or reflection of human activities and difficulties rather than as something important or valued in itself. The rural environment celebrated by pastoral writers as an escape from urban decay is also a human environment, many ecocritics would note, with agricultural lands constructed by and for humans. In addition, such literature ignores the people who occupy these pastoral landscapes or objectifies them as part of the scenery. While some ecocritics have explicitly rejected post-modern theory, post-modernist ecocritics would further reject the dichotomies assumed, shaped, and disseminated by the literature of pastoralism between city and country, between human and nature, between artificial and natural.

The study of early literatures, and their constructions of the human, of animals, and the environments built and occupied by humans, are important for an understanding of contemporary ecocritical crisis. In an ecofeminist analysis of the constructions of landscape in early Icelandic sagas, Margaret Clunies Ross points to the use of kennings in skaldic poetry that 'conceptualize the earth as an animate female being' (1998: 182 n. 4) to argue that settling Iceland 'succeeded by means that involved the symbolic or ritual expression of masculine power' (Clunies Ross 1998: 161). In an essay on textual editing in ecocritical studies, Michael P. Branch argues for the importance of early texts to our understanding of literary representations of landscape:

> If ecocritics are to construct a more complete and accurate understanding of how landscapes are understood and depicted in literature, it is essential that we broaden our thinking to imagine nature writing as a category that includes sermons, settlement, narratives, and government reports – as well as personal essays – and that we recover and examine the works of earlier writers who may be overlooked because their understanding of the natural world is predicated upon ideological or aesthetic assumptions different from our own (6).

Similarly, Hubert Zapf argues that literature, with its often deliberately non-transparent use of language, has the power to illuminate cultural blind spots about environmental problems:

> The literary works of the past appear, to a number of ecocritics, as anticipating the ecological knowledge of modern times.... Literature, from its very beginnings, has contrasted alienating structures of civilization with alternative forms of life embedded in concrete forms of a culture/nature exchange (55).

In 'The Landscape of Anglo-Saxon England,' Nicholas Howe categorizes representations of the landscape in terms of their emotional, metaphorical, or allegorical force as 'inherited, invented, [or] imagined' (Howe 2002: 91). In his discussion of landscape as 'invented' in Anglo-Saxon charters, Howe articulates an approach that resonates with ecocritical concerns, 'To invent a landscape is to order the natural terrain, or to impose organizing divisions on it, so that it becomes a human creation' (Howe 2002: 91). However, the imagined landscape remains for Howe something like traditional setting, in that he reads landscape functionally in this mode as a psychological or spiritual expression of, or as a counterweight to, human emotion. In the article, Howe focuses on the relationship between humans with respect to the land they describe. In his book *Writing the Map of Anglo-Saxon England*, Howe reads the *Beowulf* manuscript as a 'book of elsewhere' (Howe 2008: 178) that, together with the other texts in the manuscript, demonstrates the Anglo-Saxons' sense of their own place in Christian Europe and beyond.

Alfred Siewers' discussion of the *Guthlac* poems in the context of Celtic literature takes the first sustained, explicitly ecocritical, orientation to early medieval literature. He argues that landscape functions allegorically in Anglo-Saxon literary texts:

> Cultural topography of emerging Angle-land was to be found in texts of Christian Anglo-Saxon culture, in allegorized form. The new literary monumentalizing was ultimately totalizing in intent, seeking as it did to control the narrative of land, ancestry, and identity through written text in which engagement with the physical land became increasingly symbolic and relative to a more transcendent spiritual cosmos and polity. (Siewers 2003: 6)

Siewers also points out that the Anglo-Saxon myth of migration as proposed by Howe is relevant to an ecocritical analysis of the sea; in a section of the essay titled 'Into the Otherworldly Waters,' Siewers notes a similarity between Guthlac's fens and Grendel's mere, and calls the fens in which Guthlac takes up his solitary refuge 'wetlands as primeval wilderness' (Siewers 2003: 8). Bruce Holsinger, Elaine Treharne, and Sarah Kay have written about the fact that medieval texts survive on skin, in texts written out by hand, and Matt Low's essay on landscapes is a valuable ecocritical reading of Anglo-Saxon elegies.

The work of scholars of later medieval literature who have taken interest in ecocriticism is valuable in opening avenues for ecocritical thinking about Old English texts in that they also deal with the pre-modern and the ways in which philosophers, historians, and scholars of modern literatures have constructed the medieval as 'other' or have treated it as irrelevant for ecocritical analysis. Studies by Jeffrey Jerome Cohen, Karl Steel, Gillian Rudd and Susan Crane have challenged and enabled my thinking, as detailed in the chapters of this book, in some cases pointing to fruitful avenues for thinking ecologically about medieval texts, but more importantly demonstrating the value and importance of thinking through a literature that is remote in time and philosophically and artistically strange to the modern world using the insights of a contemporary critical and theoretical mode. Karl Steel's ground-breaking work very fruitfully investigates the functions of literary depictions of animals in constructing the human in later medieval literature. Gillian Rudd's *Greenery* is an excellent engagement with late Middle English texts and their literary landscapes. Jeffrey Jerome Cohen's *Animal, Vegetable, Mineral* and *Inhuman Nature* both bring together numerous valuable essays that take ecocritical approaches to medieval and Renaissance literatures.

Anglo-Saxon Texts and Ecocriticisms

This book does not attempt a comprehensive survey of ecocriticism and Anglo-Saxon texts. Ecocriticism has become so extensive a field, and there are so many Old English and Anglo-Latin poems and prose texts of interest for an environmentally-focused study, that to attempt to review them all in a single volume would allow only a superficial overview. Instead, I have in this book re-read a relatively small number of Anglo-Saxon texts from an environmentally committed point of view, bringing to bear different ecocritical perspectives in each chapter, including ecofeminism, post-colonial ecocriticism, critical animal studies, and object-oriented ontology, while discussing objects and animals as well as wilderness, ruins, and seas. In juxtaposing the utilitarian view of nature expressed in *Beowulf*, *The Ruin*, and the lives of Guthlac – Latin and Old English, prose and poetic – with the treatment of the human as the enemy and animals and objects as subjects in the Old English *Exeter Book* riddles, I wish to make an argument overall that the ideas that enabled the Industrial Revolution and the climate crisis of today were already in circulation in the Anglo-Saxon period. But Anglo-Saxon texts also articulate the ability to value nature intrinsically, to assert that humans co-exist with 'the natural world' and must live in harmony or in tension with it. We need to come to grips with the long reach of the idea that some humans have the 'right' to treat the natural world as well as other humans as 'resources,' in order effectively to challenge the hegemony of these ideas in contemporary culture.

Studying several Anglo-Saxon texts in environmental detail has revealed, broadly speaking, a contrast between *Beowulf* and the lives of *Guthlac* on the one hand, and the *Exeter Book* riddles and perhaps *The Ruin* on the other. *Beowulf* and the *Vita Guthlaci* along with the Old English versions of the *Life of Guthlac* tell different stories about the landscape and humans' relationships to it, but both sets of narratives present humans as entitled to using the world – animals, plants, landscapes, and territory occupied by 'other' humans – for their own purposes. The *Exeter Book* riddles, on the other hand, give voice to animals, plants, and objects made from ore, opening a door to the idea that the non-human possesses agency. In the chapters that follow, I take a variety of ecocritical approaches to these texts.

Chapter 2 takes an ecofeminist approach to depictions of the sea in *Beowulf* as well as in biblical epic and saints' lives. These poems about heroes, monsters, and monstrous humans suggest that humans have dominion over the earth and its creatures. In *Beowulf*, *Andreas*, *Elene*, and *Exodus*, human characters treat land and animals as 'natural resources' that are (with interesting exceptions) limitless. Grendel's mother is deeply intertwined with marginal marshland and her monstrosity reflects one aspect of cultural conflations of femininity with 'nature' in contrast to a more reasoning and more 'human' masculinity.

Chapter 3, on ruins, complicates this easy assumption of human dominion over the earth. This chapter revisits *Beowulf* and *Exodus* in addition to considering *Genesis A* and *The Ruin*. Attending to ruins and ruined dwellings in these poems makes it possible to locate locates humans in environments that are unstable, and in which they sometimes lack control, whether because of acts of God or natural decay. Ruins and ruination embed humans in the natural world rather than depicting them as separate from it or opposed to it. The lack of human builders in the Tower of Babel episode is echoed by the absence of humans who can make repairs in *The Ruin*. Conversely, the ruination of Heorot in *Beowulf* is a consequence of human presence, specifically of human violence. These Old English literary texts describe ruins in meditations on human transience and the fall and rise of different cultural formations.

Chapter 4 addresses notions of wilderness and colonization as they appear in the lives of Guthlac: Felix's Latin *Vita*, the Old English prose translation, and the verse adaptation, *Guthlac A*. Saint Guthlac is a former warrior who colonizes as his hermitage an island in the midst of the East Anglian fenland. The area is called uncultivated wasteland, despite evidence to the contrary, and the former British residents are conceived of as demons. Like the monsters of *Beowulf* and the monstrous cannibals of *Andreas*, demons can be slaughtered without compunction. The insights of post-colonial theory allow a re-reading of the versions of *Guthlac* and a coming to an understanding of colonizing ideas and ideologies, or precursors to them, as they emerged in early Anglo-Saxon England. Rather than emerging after the English began traveling to other continents, ideas about 'others' that were articulated as early as the eighth century in *Guthlac* shaped how the English viewed the people they encountered in other places – and continue, today, to enable ideologies that see some humans as less human than others.

Chapters 5 and 6 both explore the ecocritically very rich territory of the *Exeter Book* riddles. As with the volume as a whole, these chapters do not attempt a survey of all the riddles that could profitably be read with an ecocritical eye, but instead make careful and extended readings of a sampling of them. Chapter 5 examines the Riddling depiction of wild birds in the context of the formulaic depictions of beasts of battle from poems in the epic tradition, and it investigates the ways, in the *Exeter Book* riddles and related texts, that eating animals is presented as 'natural'. The chapter also examines the complex interactions of subject and object, speaker and audience, in Riddle 26, in which an animal is skinned and its hide made into a book of scripture. The process is described in the voice of the animal as the 'I' of the poem, which refers to humans as the enemy, challenging easy assumptions about human dominion or dominance over animals.

Chapter 6 turns to several riddles the solutions of which are objects made from naturally occurring sources, including trees, deer antlers, and ore from the earth. The transitions between animals and trees and the objects made from them blur the lines between sentient beings, living non-sentient organisms, and everyday things as well as ritual objects. In an on-going examination of how human and object are intertwined, and thinking through Timothy Morton's concept of 'hyperobjects' (2013a), the chapter delves into how several of the riddles reveal operations of gender and power in Anglo-Saxon society, consigning some humans to the margins while centralizing some people and elevating some objects to the point of veneration.

In the conclusion, I draw connections among the various theoretical approaches used throughout the volume and draw out intersections among them that reveal potential weaknesses in individual approaches. I suggest areas for future research, including locating traces of Anglo-Saxon environmental thinking in a longer literary and historical view and a broader geographical one. And I emphasize again the need to connect studies of literary ecologies with environmental commitments.

Throughout the volume, I argue that attention to Anglo-Saxon texts – poems and poetry, Old English and Latin – is rewarded by thinking from an ecocritical perspective. Though ecocriticism is often focused on the modern, on the present, environmental theories can also be enriched by the consideration of how texts from 1000 years ago imagine the interactions between humans and their worlds, natural and built. As I argue throughout this book, it is not the Industrial Revolution that caused us to think of our environment in terms of 'natural resources' for human consumption, but the pre-existing idea that humans could use the natural world, including other humans, in pursuit of their own needs and desires, that enabled the Industrial Revolution and subsequent commercial, colonial, and political enterprises the effects of which include environmental degradation and climate change as well as social inequalities.

2 Imagining the Sea in Secular and Religious Poetry*

Introduction

The sea figures as a major element in several Anglo-Saxon poems across different genres, including the secular epic *Beowulf*, the biblically based *Exodus*, and the saints' lives *Andreas* and *Elene*, as well as the elegiac *Seafarer* and *Wanderer*. In these poems, the sea has often been read as a setting for human or divine actions, reflecting human interests and concerns. Val Plumwood has pointed to the problems that arise from reading 'nature' (i.e. the non-human) as instrumental; such interpretations inscribe and reinforce dualistic conceptions of humans as distinct from nature (142-60). The sea of Anglo-Saxon literary and documentary texts is not simply a static stage for human actions, but a very strong presence, interacting with, influencing, and affecting the human characters. To what extent an Anglo-Saxon auditor might have imagined the sea possessing its own agency is a question worth considering.

Seascape is different from landscape because humans can travel on it, swim or dive through it, consume fish and shellfish and plant life from it, but cannot live in it. Stacy Alaimo points out that 'the open seas have long been considered empty space... the construction of the ocean in industrial capitalism has been that of a "vast void", an "empty transportation surface, beyond the space of social relations"' (Alaimo 2013: 234). But even though the sea is uninhabitable, it was of crucial importance to the Anglo-Saxons, who relied on it for food and transportation. Lawrence Buell's distinctions between 'space' and 'place' are useful for thinking about the landscapes and seascapes of *Beowulf* (2005: 63-71). 'Place' is marked with human habitation, layered with human memories, textured with information about its features and about the humans who have lived in it, and might include homes, graves, and agricultural locations. 'Space,' on the other hand, is abstract, largely unknown, lacking in personal memories, and unmarked by human activity. Wilderness is often envisioned as 'space,' devoid of human habitation, though William Cronon has pointed out that wildernesses are often ideological constructs created by 'discoverers' or other invaders who slaughter the prior inhabitants or define them as sub-human (e.g., 'savages,' 'cannibals') in order to construct a location as uninhabited (1995 passim). The sea may be space: it is more or less uninhabitable for humans, impervious (at least on the surface) to marking by human activity. Indeed, a ship may create a wake, but the water soon returns to its 'natural' status, with only the surface ruffled by wind and tide, and unplumbed depths hidden below. The same location could be simultaneously space and place, depending on who is seeing or occupying it. The sea, like a lake or marsh, might be considered a 'place' by a person intimately familiar with its geography, its contours and tides as well as the creatures that live in, on, and near it, and able to navigate safely around shoals and through strong currents. For those who have never traveled a particular part of the sea, it would be a 'space,' an abstraction unknown and undefined, and imaginable as distinct from its utility to humans.

Within literary studies of Old English texts, the sea has been seen as a lexical problem, or read either as a setting in which the action takes place or as an allegorical formulation that allows access to religious idea(l)s. Marijane Osborne reads the sea as a setting; Susana Fidalgo Monge investigates the Old English lexicon for 'sea' in *Beowulf*, *The Wanderer*, and *The Seafarer*; Phyllis Portnoy compares verbal and pictorial seas in biblical iconography; and Nicholas Howe (1989) and Ziegelmaier read the sea allegorically. Reading ecocritically interacts with and extends such interpretations, enabling a different view of the ideals, unarticulated assumptions, and preoccupations of Anglo-Saxon literary culture.

Sea Crossings: *Elene, Andreas, Exodus*

Three very different sea-crossings in Old English poems point to the wide range of literal and figurative interpretive possibilities. Elene and the Roman soldiers who accompany her enjoy an exuberant journey across the sea from Rome to Jerusalem in search of the cross on which Jesus was said to be buried. Andreas and his companions, in contrast, endure terrifying winds and waves on their sea-voyage to Mermedonia, where they are to rescue Matthew. In *Exodus*, conversely, the Israelites walk across the sea floor ahead of Egyptian warriors, who are trapped and killed by the waves crashing back over them. *Beowulf* narrates several sea-voyages as well; these will be considered separately later in the chapter. In all of these poems, including *Beowulf*, the sea functions instrumentally, serving as an index of human concerns, a metaphor for human emotions, and/or an allegorical indicator of human spirituality. In contrast to aspects of the natural word described in the *Exeter Book* riddles in chapters five and six, the sea is solely a stage for human concerns.

Numerous references across the corpus of Old English poetry to ships as 'wægflota' ('wave-floater,' *Andreas*, l. 487, *Beowulf*, l. 1907) 'sæflota' ('sea-floater,' *Andreas*, l. 381) and 'sæhengeste' ('sea-steed,' *Andreas*, l. 488) or 'sæmearh' ('sea-steed,' *Andreas*, l. 267; *Elene*, ll. 228, 245) function to normalize the sea's functionality for human transportation. A similarly utilitarian view of the sea comes across in the terms 'faroðstræt' and 'merestræt' ('sea-street,' *Andreas*, ll. 311, 898; *Elene*, l. 242). These terms conceptualize the sea as unremarkable, a surface for transportation comparable to any street passing through towns and between fields. They ignore the sea's depths and the animals and plants that dwell on or in it when they treat it as useful to humans insofar as it provides a means of transport. But references to the sea as the domain of the animals that live in it complicate this identification: The sea is also 'fisces bæð,' 'seolpæð,' and 'hranrad' ('fishes' bath,' 'seal-path,' and 'whale's road,' *Andreas*, ll. 294, 1714, and 266), as well as 'swanrad' and 'ganotes bæð' ('swan's road' and 'gannet's bath,' *Beowulf*, ll. 200, 1861). Animals living in and on the water could be used for food and skins, but these compounds also suggest that the sea, as 'stræt,' was shared with other creatures, and not simply the domain of humans. The use of 'hwales eðel' ('whale's homeland,' *Andreas*, l. 274) raises the possibility that the Anglo-Saxons recognized that they passed across it and not through it, as visitors rather than inhabitants, and viewed the sea as the rightful domain of whales, and not of humans. The idea that animals, as well as plants and even stones, could claim space in or on the earth and regard humans as enemies competing for it, is explored in detail in chapters five and six, on the *Exeter Book* riddles. The idea that humans shared the ocean with its creatures, and even that whales and other animals were its legitimate inhabitants, gains credence in juxtaposition with the ideas expressed in the *Exeter Book* riddles.

Depictions of the sea in *Beowulf* and *Elene* articulate pleasure in the environment. In *Elene*, the narrator describes the slap of sea on the hull of a ship as an unthreatening event: 'Bord oft onfeng / ofer earhgeblond yða swengas; / sæ swinsade' ('A plank often received, over the waves' surges, the blows of the waves; the sea sang out,' ll. 238b-40a). The sound of waves striking the ship's planks in the passage is an occasion not for fear of the power of the sea to harm, but for confidence in the strength of the ship and delight in traveling over the water. The ship plays in the sea and subdues the power of the waves: 'þær meahte gesion, se ðone sið beheold, / brecan ofer bæðweg, brimwudu snyrgan / under swellingum, sæmearh plegean, / wadan wægflotan' ('He who beheld that journey might see there the sea-wood hurrying, taming the water-way under the swellings, the sea-horse playing, wave-floater wading,' ll. 243-46a). The poetic play of language, the ship conceptualized as a horse prancing on the sea, evokes the joy of traveling on a ship that, in the hands of a competent captain and crew, plies waves large enough for aesthetic, possibly awed appreciation, but not so large as to threaten the ship. (All translations are my own, except as noted.)

The Latin analogues have none of this celebration of the sea; the source for *Elene*, the *Acta Quiriaci*, states only briefly: 'When Constantine had learned from them [his teachers] where the Lord had been crucified, he sent Helena, his mother, to seek the holy wood of the Lord's Cross and to build a church in the same place' (*Sources and Analogues* 61). In *Beowulf*, a similar, apparently aesthetic observation of seafaring appears in the comment that, as Beowulf and his men board their ship, 'streamas wundon, / sund wið sande' ('sea-streams twisted, water with sand,' ll. 212b-13a; all references to the text of *Beowulf* follow Klaeber's edition). Beowulf's ship is also compared to a bird: 'Gewat þa ofer wægholm, winde gefysed, / flota famiheals fugle gelicost' ('Then, speeded by the wind, the foamy-throated floater went over the sea, most like a bird,' ll. 217-18). In *Elene*, the ship cavorts in the water like a horse crossing a shallow ford; in *Beowulf*, it flies over it.

From an ecocritical point of view, such articulation of pleasure in the movement of a ship on the sea is complex. It may be viewed as obfuscating human exploitation of the sea's resources, though the fishing industry and the transportation methods used in the Anglo-Saxon period did not have impacts as severely destructive as such activities when they are employed today alongside other appropriations of natural resources from the oceans. The expression of joy in journeying on the sea, however, might be read as celebrating, on its own terms, a landscape in which humans cannot build or live, and unstructured by human intervention. Unlike a pastoral idyll, such a view of the sea does not depend for its idealized force upon opposition to a squalid urban or otherwise problematic earthbound dwelling, though it still contrasts with and presupposes landed human habitation. Such celebration of the sea remains anthropocentric, in that it does not afford to the sea any ethical imperative on its own terms.

In contrast to these celebrations of joyous sea-travel in *Beowulf* and *Elene*, poems as different as *The Seafarer* and *Andreas* depict the sea, and travel across it, as cold, miserable, and terrifying. When God instructs Andreas to rescue Matthew before he is to be killed three days hence, Andrew demurs, asking how he can travel from Achaia to Mermedonia in just three days. He refers specifically to the difficulties of ocean travel, calling the sea 'deop' ('deep,' l. 190) and 'wæterbroga' ('frightful flood,' l. 197) and adding, 'ne me herestræta / ofer cald wæter cuðe sindon' ('nor are the highways over cold water known to me,' ll. 200b-01). Once he is underway, 'hornfisc plegode, / glad geond garsecg' ('the garfish played gladly throughout the sea,' ll. 370b-71a), but the gathering storm troubles Andrew and his companions because they do not know that they are being ferried by Jesus in the company of two angels:

> Wedercandel swearc,
> windas weoxon, wægas grundon,
> streamas styredon, strengas gurron,
> wædo gewætte. Wæteregsa stod
> þreata þryðum.
>
> The sun darkened, winds grew powerful, waves gnashed, streams stirred, ropes grated, sails grew wet. Water-terror arose despite the might of the troop. (ll. 372b-75a)

This storm is suggested by the Latin analogue for *Andreas*, according to which 'ita enim insurrexerat illis validissima tempestate maris, et fluctuum' ('a great tempest of sea and surge had risen against them,' *Acta Andreae* 47; *Sources and Analogues* 198), but is greatly embellished in the Old English text.

The contrast between cavorting fish and frightened men is interesting in that it provides a momentarily altered perspective, a brief acknowledgment of animal agency, in a poem otherwise rather relentlessly about humans – or rather, it must be acknowledged, about men. The slippage in the instrumental view of the sea provides a momentary recognition that perspectives other than the human exist, and matter. Perhaps an Anglo-Saxon view of nature in contrast with that of the Latin source has intruded, only to be superseded immediately by the primacy of the human point of view. The joyous sea-crossing of *Elene* could be interpreted as a figuration for baptism or read as a journey enabled by divine favor and therefore experienced as delightful. Such readings do not supplant, but must be read alongside, the more directly celebratory tone of the passage. In the case of *Andreas*, however, this storm is heavily metaphorical, a direct consequence of Andrew's lack of faith in God's power to facilitate his mission to the Mermedonians, indicated also by his failure to recognize Jesus.

God suggests that Andrew's frightened thanes leave him behind, but they refuse to do so, setting a better example than Andrew, who had initially said he could not rescue his fellow apostle from being consumed by cannibals. In rejecting the possibility of making the journey, Andreas also mentions the wide lands he would have to cross in order to arrive in Mermedonia within three days, but the subsequent narration of his journey does not include any description of travel on land. The focus is on the difficulty of the stormy, frightful, and strange sea journey, in sharp contrast to the solid ground where Andreas and his men arrive, to their relief.

The island is also beset by watery storms: Andreas's imprisonment is made harsher by a night of severe winter weather, with frozen water in various forms:

> Snaw eorðan band
> wintergeworpum. Weder coledon
> heardum hægelscurum, swylce hrim ond forst,
> hare hildstapan, hæleða eðel
> lucon, leoda gesetu. Land wæron freorig
> cealdum cylegicelum, clang wæteres þrym
> ofer eastreamas, is brycgade
> blæce brimrade.
>
> Snow bound the earth in winter storms. The weather grew cold with hard hail-showers and with rime and frost, hard warriors; locked up noble fighters and people's homes. The lands were frozen with cold icicles, water's force shrank over streams of water, ice bridged black sea-roads. (ll. 1255a-62a)

This storm seems not to be a feature of the earthly environment, but something external to it. Karin Olsen argues that in *Andreas* the sea is described as a weapon threatening the human band: 'The presentation of Andreas's and his comrades' past struggles on a stormy sea contains all the commotion of an attack by a personified relentless aggressor upon the defenseless, terrified sailors' (387). Lindy Brady finds that the sea voyage possesses a 'characteristically Anglo-Saxon tenor ... with its similarity to a Viking raid' (671). The first three riddles of the *Exeter Book* similarly describe storms as attacking land and its human occupants as earthly phenomena, not divine punishment, as discussed in greater detail in chapter 6. The weather is treated as a warrior, an 'avenger' causing buildings to burn and men to die. The storm, as narrator of Riddle 3, describes itself in opposition to 'my lord' and attacks buildings and ships as well as the sea. In the riddles, the storms destroy humans and human dwellings as well as attacking land and water, in contrast to *Andreas*, where the force of the sea is directed specifically at Andreas's ship, and at Andreas and his men.

The sea of *Elene* and *Beowulf* is benign. The contrasting presentation of sea as distinctive from land in *Andreas* gives it a quality of alterity, what

Lawrence Buell defines as 'space' as distinct from 'place' (2005: 63-71). 'Place,' in Buell's definition, drawn from Yi-Fu Tuan's work on cultural geography, is that with which humans have connection, whether because it is personally known or because there is an ancestral or emotional connection. 'Space,' by contrast, is unknown, foreign terrain. The island of the Mermedonians is 'place' to its inhabitants and would seem to require non-recognition as 'space' for Andrew and his companions, yet in its presentation in contrast to the sea it seems even to them to be more like 'place.' Lindy Brady has, moreover, argued that the description of Mermedonia has been given familiar details drawn from the actual fenlands of England. On water, Andrew was at the mercy of God, ferried by Jesus across a stormy sea with the capacity to kill. On land, Andrew will ultimately exercise control over the demons that plague him: he will wield speech, resist torture, rescue Matthew, and slaughter numerous Mermedonians by causing water to gush violently forth from an old stone made by giants.

Importantly, from an ecocritical point of view, both land and sea are for the poem simply setting, functioning instrumentally and metaphorically subordinated to human concerns. The description of the violent sea-crossing is interrupted so that Andrew can recite material from the gospels, prompted by Jesus himself, who turns out to be the ship's navigator. The descriptions of the landscape are distantly secondary to the actions that occur on its surfaces. For the poet, the sea is utilitarian in terms of its metaphorical relationship to Andrew's state of mind as he addresses Jesus in his uncertainty about undertaking the voyage to rescue Matthew, and is then calmed by the recitation of the gospel narrative. From the point of view of Andrew as a character, the sea is utilitarian in terms of being available for transit. In addition, land is utilitarian in being noticeable for its distinction from sea – it can be crossed without the need of a ship or other conveyance – and in terms of providing, just at the right moment, an age-old stone, 'eald enta geweorc' ('the ancient work of giants,' l. 1495), from which Andrew is able to command water to gush forth and kill. The poet takes for granted that land and its non-human inhabitants exist only for the sake of humans. This is not a Christianity of stewardship, of taking responsibility for the earth and its creatures and managing them, but a world-view instead that takes humanity in opposition to nature and controls it (sometimes explicitly through God) for human purposes. The movement across the sea also functions in utilitarian fashion to provide a physical analogue for Andrew's transition from reluctance to journey to Matthew, to understanding that he must do so, and finally undertaking the project despite being tortured by the Mermedonians.

In *The Seafarer*, the wretched potential of sea-travel functions metaphorically for exile from human company rather than for lack of faith. In the poem, 'bitre breostceare' ('bitter breast-cares,' l. 4) are paralleled by the miseries of winter sea-travel: 'Calde geþrungen / wæron mine fet, forste gebunden, / caldum clommum' ('Pinched with cold were my feet, bound with frost, with cold fetters,' ll. 13b-15a). The sea is also place of exile in *The Wanderer* and, as in *The Seafarer*, it is full of wintry torments. The 'hrimcealde sæ' ('rime-cold sea,' l. 4) beset by 'hrim ond snaw, hagle gemenged' ('frost and snow, mixed with hail,' l. 48) reminds the Wanderer of the absence of 'freomægum' ('free kinsmen,' l. 21) and other company 'in meoduhealle' ('in the mead hall,' l. 27). Likewise, the 'cald wæter' ('cold water,' l. 201a) of *Andreas* or the 'sincalda sæ' ('perpetually cold sea,' l. 473a) of *Exodus* reflect an Anglo-Saxon reality that the oceans around the British Isles are very cold. The poet of *Exodus* apparently did not know about the Red Sea's warm temperatures. The fictional island of Mermedonia is also located in a perpetually cold sea.

The sea-crossing in the Old English *Exodus* contrasts markedly with those in *Elene* and *Andreas*. Like the *Andreas* sea-crossing, that in *Exodus* is characterized by divine intervention, but the similarity ends there. Andrew and his companions cross the sea in a boat piloted by Jesus, frightened and threatened but ultimately unharmed by the storm. In *Exodus*, the Israelites cross the Red Sea by walking across the bottom, and God then causes the water to rush back, slaughtering the pursuing Egyptians.

The biblical Exodus includes the detail that 'the waters were indeed just like a wall' ('erat enim aqua quasi murus,' Exod. 14: 22). When the Israelites have crossed the sea and the Egyptians remain behind them, 'reversaeque sunt aquae' ('the waters returned,' Exod. 14: 28). The Old English poem expands on this description, depicting the water as a fortification, referring to it as 'wealfæsten' ('fortress,' l. 283), 'staðolas' ('firmaments,' l. 285), 'randgebeorh' ('shield-wall,' l. 296). As the water crashes down on the Egyptians in pursuit similar language is used: 'weallfæsten,' (l. 484) and 'randbyrig' (l. 464) are repeated, and 'meretorras' ('sea-towers,' l. 485) added. (This passage is discussed in greater detail in Chapter 3, 'Ruined Landscapes.')

In the context of such varied depictions of the sea in *Elene* and *Andreas*, these lines provide a further example of the natural world being depicted in instrumental terms in a narrative about human concerns, rather than with any sense of the importance of the sea for its own sake. The sea becomes an instrument of divine intervention in human history. Rather than the celebration of the sea in *Elene*, the sea of *Exodus* is imagined as a tool of God put to the service of humans as savior of the Israelites (and typologically, of future Christians) and subsequently as killer of Egyptians. Lynn White has argued that a Christian conception of the earth and its creatures as given to humans to support their lives opened a door to the abuse and exploitation of natural resources, in opposition to European paganism as well as the religions of Native Americans, in which humans have the responsibility to tend to the earth for the sake of future generations (3-14). Whether the context in which such ideas are expressed in these Anglo-Saxon poems is unique to Christianity or also a survival of Anglo-Saxon paganism is probably impossible to ascertain, though it is interesting that they appear not only in texts transmitted into Anglo-Saxon England through literate Christianity, but also in an oral-formulaic epic of pagan, Germanic origins. Christianity has also been credited with developing a notion of stewardship for the earth, but these scenes in which the sea is imaged as a divine tool and commentary upon human affairs point more to the idea of the earth as placed at the service of humans rather than of humans as guardians for the earth and its resources. Such ideologies are not post-medieval inventions. They are not symptoms of the empirical expansion or Industrial Revolution that occurred hundreds of years later, but rather underlying social attitudes that helped to enable both to occur.

Beowulf and the Sea-Creatures

As noted above, the sea in *Beowulf* is established early on as space: it is 'hron-rad' ('whale-road,' l. 10) and 'swan-rad' ('swan's path,' l. 200), a space to cross in a ship most like a bird ('fugle gelicost,' l. 218). Beowulf and his men travel across a sea-space in order to arrive at Heorot, and they travel the same sea-space to get back home, but there is no sense that the journey is enabled by a specialized knowledge of navigating that particular sea. Building, as does Buell, on the work of cultural geographer Yi-Fu Tuan, Gillian Overing and Marijane Osborne note that landscapes are created by the imagination: 'The terrain of place is then substantially internal, the picture made within the frame of individual perception' (xxi). The sea depicted here is the imagined sea of the author and/or scribe of *Beowulf* rather than the observed sea of any actual traveler.

The sea through which Beowulf has travelled in his contest with/against Breca is another space, undifferentiated for either of the swimmers by local knowledge of its particular characteristics. The importance of this episode is signaled by the fact that it is narrated twice, first by the mocking Unferth, and then in a revised version accepted by the Danish audience, in which Beowulf claims that he was the stronger swimmer. Beowulf tells the company he had stayed by Breca's side until they were separated by wind and waves as they struggled against 'wado weallende, wedera cealdost / nipende niht, ond norþan wind' ('rolling waves, the coldest of weather, the falling of night and wind from the north,' ll. 546-47).

The status of this sea as space is further suggested by the battles between Beowulf and its 'native' or 'natural' inhabitants. These are not, however, referred to as fish or birds, but rather 'niceras' ('sea monsters,' ll. 422, 575). The term appears elsewhere only in *Alexander's Letter to Aristotle*, where Andy Orchard translates it as 'water-monsters [hippopotami],' in a nod to the Latin text of the *Letter*, which has 'hippopotami' at this point (209, 234-35). For the Anglo-Saxon author, scribe, and audience, then, these creatures are generically monstrous, rather than known animals of any kind. They are named with the same word used for the creatures that inhabit Grendel's mother's mere until the moment of her death (l. 845), creatures that are also called 'wyrmas ond wildeor' ('serpents and wild animals,' l. 1430) as well as 'sellice sædracan' ('unusual sea-dragons,' l. 1426). The latter term links them with the dragon or 'wyrm' (e.g. l. 2567) that Beowulf later kills, yet it also suggests perhaps eels – toothy, aggressive fish that could be terrifying indeed to fight while swimming in the water with them. The presence of eels in the rivers and marshes of England is attested by the name of the island town of Ely in East Anglia, surrounded during the Anglo-Saxon period by marshland and so named, Bede tells us, because the marshes were populated with 'copia anguillarum' ('large quantities of eels,' 4.19: 199, translated in the Old English version as 'genihtsumnesse ælo,' 'huge quantities of eels,' 324, l. 8).

The use of the generic term 'niceras' to name the creatures Beowulf and Breca encounter during their swimming-match, rather than a term for any specific type of animal, makes the sea strange, and this estrangement likewise renders the sea 'space' rather than 'place.' Ultimately, Beowulf kills nine sea-monsters so that, he says, the sea will thereafter be safe for travelers: 'syðþan na / ymb brontne ford brimliðende / lade ne letton' ('afterwards never in the steep seas would they hinder the passage of sea-voyagers,' ll. 567b-69a). Whether or not they have voices, and whether such voices could be heard or understood by humans, the creatures Beowulf slaughters in his sea-journey are sentient beings. They experience his entry into the sea, bearing arms and wearing armor, as an intrusion into their home, which they inhabit as 'place.' Beowulf, conversely, enters it as an intruder; he nevertheless formulates their dwelling in this 'space' as illegitimate and in need of correction. Beowulf re-inscribes this seascape as a different kind of 'space,' one that is safe for his fellow humans to travel on because he has slaughtered its original and legitimate inhabitants.

Beowulf provides justification for the slaughter of these sea-creatures, much as he feels the need to justify his other conquests. In his negative account of the events, Unferth does not do so; this distinction suggests that in order for Beowulf to present himself positively he needs to volunteer just cause for the killing of the creatures. It is perhaps the fact that Beowulf kills the sea-creatures where they live, much as Grendel kills the Danes in their dwelling-place, that requires explanation. The ecological theorist Christa Grewe-Volpp points to the problematic nature of speaking 'for' the natural world. Beowulf's justification of the death of the creatures from the sea suggests they possess agency, yet Beowulf appropriates any such agency by speaking for them and creating a narrative of their illegitimacy.

Beowulf states that he has never heard of another man surviving such an ordeal (though, in fact, Breca did). This journey through watery depths, from which Beowulf is lifted by the waves onto the shore of Finland, could be interpreted as a metaphorical (re)birth into manhood. This journey begins, it appears, with a childish dare from a youthful friend, but ends with Beowulf having provided his community a service in a feat of daring sufficiently courageous to be known among the Danes across the sea before his arrival. Moreover, he has made an impact on the sea itself. It is not turned into a habitation or agricultural land, but into a safer road – still apparently a 'space' to be traversed rather than a 'place' to be occupied, yet subject to human intervention that characterizes as monstrous the 'natural' or original occupants to help justify their slaughter, for human convenience.

Marsh in *Beowulf*

The sea is generally a space that humans travel on top of, using it as a 'road,' rather than swimming in or through; Beowulf is quite unusual in his ability to spend inordinate amounts of time in the water, both during his boyhood swim with Breca and later, when he will swim at length to reach Grendel's mother's underwater cavern. Kelley Wickham-Crowley observes that water 'was an integral part of the Anglo-Saxon perception of "landscape"' (85). Much of that watery landscape, however, took the form of marshes and fens, much like the terrain inhabited by Grendel and his mother (86-87). While his earlier sea encounter is undertaken in the company of Breca, Beowulf travels alone into Grendel's mother's swampy mere. The fact that Beowulf is able to survive in the mere – which is neither wholly water nor wholly land – emphasizes his status as different from other humans in the poem. This incident is described by the poem's narrator; when Beowulf later retells the episode to Hygelac, the swim through the water is nearly elided; Beowulf reduces the description of his passage through the water to a word or two (ll. 2135-36). The sea-creatures, whose deaths Beowulf felt needed justification when he argued with Unferth over the swimming-contest with Breca, are in this retelling ignored altogether. Beowulf's status, along with the monsters of the poem, as 'aglæca' ('fighter, monster,' ll. 1512, 2592) – something uncannily superhuman – is suggested by the fact that he makes two major journeys into and through water, rather than on it in a boat, and lives. The first use of this word to refer to Beowulf occurs when he is swimming down through the lake/swamp in search of Grendel's mother's lair (l. 1512).

Scholars typically call the place where Grendel's mother lives 'Grendel's mere,' but it would be more accurate (albeit more cumbersome) to call it 'Grendel's mother's mere.' Grendel attacks on land. He is said to bring his dead victims back to the lair; Beowulf says that if he loses the fight with Grendel, the monster will bring Beowulf's dead body back home to his lair to feast on. But the poem places Grendel in the cave only after Beowulf has torn off his arm and he has retreated there to die, and at second hand, as narrated by Beowulf to Hygelac (l. 2099). It is only after Grendel's death that the audience learns, along with Beowulf, that Grendel and his mother have been seen to haunt the marches, the edges of Hrothgar's kingdom, the boundaries between dwelling and wilderness, between place and space. The monsters' dwelling-place, moreover, is a swamp: a space/place that is, much like the mere, neither wholly land nor wholly water. It is bounded by forest, accessed via paths created perhaps by wild animals, perhaps by humans, perhaps by the Grendelkin themselves. Traversing swampland is possible, but difficult and potentially dangerous, requiring a level of familiarity with the specific terrain, as demonstrated by the fact that Guthlac needs a guide to show him the place where he will locate his hermitage (see Chapter 4).

In *Postmodern Wetlands*, Rod Giblett argues that swamps have been identified across cultures and times with danger, death, disease, and sin, an association born out in *Beowulf.* In various Anglo-Saxon texts, 'interpenetrations of water and land' (Wickham-Crowley 87) are occupied by demons, like Guthlac's hermitage on a fen-surrounded island, or by water-monsters such as those infesting (or living in, depending on perspective) Grendel's mere, and water needs frequently to be '"disinfected" of spirits' (Dendle 192). Grendel and his mother are twice identified as descendants of Cain (*Beowulf* ll. 107, 1261), doomed to wander for generations in sin. The horrors of the mere are described at two different points in the poem. When Hrothgar describes Grendel's mother's mere to Beowulf, he tells him: 'No þæs frod leofað / gumena bearna þæt þone grund wite' ('No one old enough lives among the sons of men who has known the bottom [of that mere],' ll. 1366a-67). Hrothgar goes on to tell Beowulf that the swamp is so frightening to animals as well as humans that a deer driven by hunters will stop, turn, and face death rather than enter its swampy ground: 'nis þæt heoru stow' ('that is not a pleasant place,' l. 1372b). Later, Beowulf approaches the place, accompanied by his own men, as well as by a retinue of Hrothgar's fighters. The men find Aeschere's head, which Grendel's mother has left to mark the entrance to her home, an act that parallels the display of Grendel's arm in Heorot.

The waters surrounding Grendel's mother's cave are riddled with swimming monsters. One of Beowulf's men idly looses an arrow at one of these monsters as it breaks the surface, and, as it flounders, several of the warriors stab it to death. The poem calls the beast a 'wæg-bora,' which Seamus Heaney translates as 'lake-birth' (l. 1440), following a definition proposed by Friedrich Klaeber (glossary, *s.v.*). The compound is a difficult one; with 'bora' derived possibly from 'borian' (bore, drill), and possibly from 'beran' (to carry, to be borne, to give birth to, to beget). The first part of the compound, 'wæg,' can be translated more straight-forwardly as 'wave, water, sea, billow.' The compound could refer to the sea-creature as that which bores into or through the waves, or perhaps that which is carried by the waves. Heaney's suggestion that the sea-beast is birthed by the water is supported by the strong link between Grendel's mother and the sea-creatures.

Meanwhile, Beowulf arms himself beside the water, already boiling with blood – perhaps Aeschere's, perhaps that of the dying animal. Following Kristeva's discussion of menstrual blood as pollution, this appears as a metaphorically polluting blood, associating the mere with the feminine, with menstruation and birth (71). Beowulf plunges into the water; it takes 'hwil dæges' ('the space of a day,' l. 1495b) before he reaches the bottom. The long, bloody, watery space through which Beowulf passes might be read as a kind of *vagina dentata*, monstrously fearsome in its geographical formulation, the toothy animals reaching out to grab Beowulf as he descends. It might, however, be argued that Beowulf in fact is able to reach the bottom of the mere only because Grendel's mother grabs him and carries him down with her to the bottom (ll. 1506-07).

Jane Chance reads the ensuing fight in sexual terms, with Beowulf and Grendel's mother alternately sitting astride one another, before Beowulf succeeds in penetrating his opponent. His own sword fails, so he must use one he has found in the mere to cut off her head so that blood drips from the blade (253-54). Beowulf investigates the cavern and finds the corpse of Grendel, and beheads it too, at which point the blade melts post-coitally away, leaving only the gem-encrusted hilt. The men waiting next to the water see blood; Hrothgar's thegns assume it must be Beowulf's and return to Heorot, but the members of Beowulf's own retinue refuse to give up hope. As soon as Grendel's mother is dead, the sea-creatures mysteriously vanish from the water, and Beowulf is able to swim unimpeded to the surface, carrying as his two tokens of the battle Grendel's head and the engraved, gilded hilt of the ancient sword.

While Beowulf's slaughter of the sea-beasts during his earlier contest with Breca was justified by the need for the Geats to journey over the waters safely, no rationale is given for the killing of one of the animals before Beowulf entered the water of Grendel's mother's mere. Their disappearance on the death of Grendel's mother suggests that the very existence of these beasts is allegorical: they are 'real' within the terms of the narrative as long as Grendel's mother lives, but as soon as she dies, they cease to have an existence separate from hers. Nevertheless, Alfred Siewers suggests that the killing of such creatures, as well as of Grendel and his mother, is problematic: 'Guthlac's exorcism of the Fens parallels readings of Beowulf's foray into the Grendelcyn's mere as the exorcism of an earlier indigenous culture' (2003: 9). As he did for the earlier seascape by killing nine 'niceras,' Beowulf re-inscribes the space in which Grendel and his mother lived: once womb/dwelling, it is now tomb.

This detail further establishes the inextricability of Grendel's mother's identity from the marsh through which Beowulf must travel to reach her cavern. The bloody passage through which Beowulf travels – from cave, through waters, and back to land – might be read as a second re-birth for Beowulf, an emergence from an earthly tomb/womb back into life, with his final release from the bloody depths enabled and aided by his waiting retainers. In a review of the history of *Beowulf* scholarship, Irving notes that previous generations of scholars have seen Beowulf as a Jesus figure; in such readings, Beowulf's emergence from what might have become his own tomb is interpreted as a kind of divine rebirth (175-92 passim). Yet the association of Grendel's mother with the mere suggests instead, or simultaneously, a re-birth from a feminized, liminal marsh, neither land nor sea, into a masculine social order, in which Beowulf's position will now be far different from what it was before his initial encounter with Grendel. Haruko Momma reads the first half of *Beowulf* as 'an evolutionary *Bildungsroman*' in which the hero seeks 'guidance to make the transition from a monster-fighting saviour to a warrior in the leisure class' (169). After the battle with Grendel, Hrothgar wished (though Wealhtheow vetoed it) to make Beowulf his heir; after this battle, however, Hrothgar issues to Beowulf a warning about the abuse of his clearly considerable powers. Beowulf has become superhuman, and perhaps rather frightening.

Ecofeminism and the Other

Ecofeminist philosophers such as Val Plumwood and Judith Chelius Stark have traced to ancient Greek culture, and thence forward into our own, a dualistic way of thinking that opposes humans to nature. Like many ecocritics, Plumwood and Stark leap from the ancients to the Renaissance without any discussion of the Middle Ages. Plumwood (2003) points out that in the *Timaeus*, Plato associates 'nature' with the human body (as distinct from reason or soul) and the emotions with the senses, with animals, landscape, and wilderness, with the feminine and with reproduction, and with chaos and the world as a place of change (80). Lanfranc, the eleventh-century archbishop of Canterbury, wrote comments on translations of Plato's *Timaeus* indicating that the text was (at that late date) known in Anglo-Saxon England, but Gneuss's *Handlist* suggests that no copy survives in Latin or Old English (50). However, as Stark argues, Augustine 'is one of the major architects who forged the synthesis of Platonism and Western Christianity in late Antiquity' (22). The works and ideas of Augustine were well-known in Anglo-Saxon England. Stark argues that Augustine accepted the Platonic division between soul and body, and she contends further that 'just as he accepted this dualistic view of human beings, he also accepted the ordering and hierarchy that gave prominence to the soul over the body and, in general, the superiority of the spiritual over the material' (22-23). Furthermore, 'Augustine relied heavily on the language of hierarchy, control, and subordination of women to men' (28-29).

The notion that humans are opposed to and elevated above 'nature' – meaning the non-human natural world – made its way to Anglo-Saxon England by way of Boethius' *Consolation of Philosophy*, which was well known in Anglo-Saxon England, surviving, according to Gneuss (151, 158), Ker (519) and Lapidge (293-94) in several Latin manuscripts as well as three copies of the Old English translation attributed to King Alfred. According to the Old English *Boethius*:

> For þi ic cwæð þæt sio sawul wære þriofeald forþam þe uðwitan secgað þæt hio hæbbe þrio gecynd. An ðara gecynda is þæt heo bið wilnigende, oðer þæt hio bið irsiende, þridde þæt hio bið gesceadwis. Twa þara gecyndu habbað netenu swa same swa men; oðer þara is willnung, oðer is irsung. Ac se mon ana hæfð gesceadwisnesse, nalles nan oðru gesceaft; forði he hæfð oferþungen ealle þa eorðlican gesceafta mid geðeahte and mid andgite.
> I said that the soul was threefold, because philosophers say that it has three natures. One of those natures is concupiscible, the second irascible, the third rational. Animals have two of these natures, like men: one of those is concupiscence, the other is anger. But man alone has reason, not the other creatures, and so he has surpassed all the earthly creatures with thought and understanding (I. 317, ll. 217-33, trans. Godden and Irvine 2:53.)

The term 'gecynd,' translated here as 'nature,' does not refer to the natural world, which is rendered instead by 'ealle þa eorðlican gesceafta' ('all of earth's creations'), which are set collectively against 'se mon ana' ('man alone'). From Boethius, the idea of humans as separate and distinct from nature, and the justification for human dominion over the earth and its creatures, makes its way into Ælfric's sermon on the Nativity of Jesus (Godden n.p.). Ælfric draws fairly closely on the language of the Old English Boethius:

> Uþwytan sæcgað þæt þære sawle gecynd is ðryfeald. An dæl is on hire gewylnigend-lic, oðer yrsigendlic, þrydde gesceadwislic. Twægen þissera dæla habbað deor and nytenu mid us, þæt is gewylnunge and yrre. Se man ana hæfð gescead and ræd and andgit.... Ðuruh þæt gescead ana we synd sælran þonne þa unge-sceadwysan nytenu.
> Philosophers say that the soul's nature is threefold. One part in it is capable of desire, the second capable of emotion, the third capable of reason. Two of these parts, animals and beast have along with us; those are desire and anger. Man alone has reason and speech and understanding... Through reason alone are we better than the irrational beasts. (Skeat I.16, I. 18, ll. 96-100 and 148-49.)

Precisely where Grendel and his mother might fall within this paradigm that would distinguish human beings from all other aspects of creation, and whether or not they possess any capacity for reason, are difficult problems. As noted above, the Grendelkin belong to the race of monsters descended from Cain (*Beowulf*, l. 107). Grendel fights without weapons, bursting in on Heorot and feasting on warriors. His mother, on the other hand, is stealthy, taking Hrothgar's favorite thegn in exchange for her son with an apparent understanding of human codes of vengeance. Katherine O'Brien O'Keeffe suggests that Grendel and Beowulf both approach 'the limits of the human' in their battle, with Grendel entering human territory and Beowulf discarding human armor (1981: 484-94 passim). Grendel's mother seems more human than her son: she wears armor and wields weapons. However, both Grendel and his mother are also described as 'ellorgæstas' ('alien spirits,' l. 1349a).

Of the two, Grendel is better known to the people of Heorot; he actually enters the hall to slaughter Danes, year after year. Conversely, until Beowulf seeks her out, Grendel's mother has only been seen from a distance; she is described as a 'mearcstapa' ('boundary-walker,' l. 1348). When Grendel's mother approaches the hall for the first time, the narrator states: 'Wæs se gryre læssa / efne swa micle swa bið mægþa cræft, / wiggryre wifes, be wæpnedmen' ('The terror was less, by just so much as is the strength of a maiden, the martial terror of a woman, in comparison to that of a weaponed man,' ll. 1282b-84). She is identified only as 'Grendel's mother,' and is given no name of her own. This detail links her with many other female figures throughout the Old English corpus, both literary and documentary, who are also nameless, identified only as someone's wife or daughter. When Hrothgar tells his coast guard that he knows Beowulf, he names Beowulf's father, but identifies his mother only as the daughter of Hrethel: 'wæs his ealdfæder Ecgþeo haten, / ðæm to ham forgeaf Hreþel Geata / angan dohtor' ('his father was called Ecgtheow, to whose home Hrethel the Geat gave his own daughter' (ll. 373-75a). Near the end of the poem, a nameless woman mourns Beowulf at his funeral: 'swylce giomorgyd (Ge)at(isc) meowle / (æfter Biowulfe b)undenheorde / (sang) sorgcearig' ('also, a Geatish woman with hair bound up sang a mournful lament for Beowulf, a sorrowful song,' ll. 3150-52a). (The passage is damaged and difficult to translate, and has been reconstructed differently by different editors, but the absence of any name or other identification is at least clear. This version is from *Klaeber's Beowulf* 270 n. 3150).

The mother is, for Kristeva, both subject and object, that which must be rejected in order to enter language (32, 40-41). Paul Acker draws on Kristeva in identifying the maternal, in the case of Grendel's mother, with horror. Grendel's mother is the abject, neither subject nor object, 'the jettisoned object [which] is radically excluded and draws me toward the place where meaning collapses' (Acker 2). Kristeva refers to the acquisition of speech as 'devouring language' (Kristeva 40); the speaking person becomes a 'fortified castle' (46). Grendel and his mother enter the fortified castle of Heorot and instead of devouring language, they consume humans. The idea of 'wiggryre wifes' ('the war-terror of a woman,' l. 1285), a female monster who has killed and apparently eaten a human male, is almost too much for Hrothgar to articulate, what Kristeva calls 'a reality that, if I acknowledge it, annihilates me' (2).

Kristeva's use of the phrase 'murky waters' (59) in a section heading in her discussion of the nature of femininity and the abject provides an uncanny echo of the association of Grendel's mother with the swamp. Grendel's mother and the water she occupies bear some similarity to the description of the storm in the *Exeter Book*, Riddle 2. The riddling speaker/narrator called the sea 'hwælmere' ('whale-mere,' l. 5), a term which evokes the references to Grendel's mother's dwelling as 'mere' (ll. 845, 855, 1362, 1603) and of Grendel's mother as 'merewif' ('water-woman,' l. 1518). Interestingly in terms of the history of scholarship about Grendel's mother, Bosworth-Toller provide semantically neutral translations for a number of 'mere' compounds, translating 'mere-fara' as 'sea-farer,' and 'mere-liðende' as 'sea-faring person.' However, they render 'merewif' as 'water-witch,' a translation followed by Klaeber and retained by Fulk, Bjork, and Niles in their revision of Klaeber's edition. In Heaney's translation, Grendel's mother becomes 'that swamp-thing from hell' (l. 1518).

Associating the sea with whales via the compound 'hwælmere' suggests enormity, even monstrosity analogous to the 'wyrm-cynn' ('worm-kind,' l. 1425b) and 'sæ-draca' ('sea-dragon,' l. 1426a) that appear alongside Grendel's mother in *Beowulf*. In both poems, the water has the feel of something that is not only occupied by living beings but also alive in its own right. In Grendel's mother's mere, 'wæter under stod / dreorig ond gedrefed' ('water rose up below, / dreary and disturbed,' ll. 1416b-17a). Similarly, in the *Exeter Book*, Riddle 2:

hwælmere hlimmeð, hlude grimmeð,
streamas staþu beatað, stundum weorpaþ
on stealc hleoþa stane ond sonde,
ware ond wæge, þonne ic winnende,
holmmægne biþeaht, hrusan styrge,
side sægrundas.

The sea [whale-mere] roars, rages loudly, currents beat against the shore, fiercely flings stone and water, seaweed and waves, against the steep cliff, when I, raging, hidden under the force of the waves, extensively stir up the earth of the sea-bottom. (ll. 5-10a)

Grendel's mother is not simply an inhabitant of the mere that she occupies; she is also a part of it. She and Grendel walk the marches; he is 'in weres wæstmum' ('in the form of a man,' l. 1352) and she is 'idese onlicnæs' ('in the likeness of a [noble] woman,' l. 1351). Both Grendel and his mother live in a kind of exile from Heorot and its inhabitants. As Stacy Klein has observed, depictions of exile in Old English poetry operate differently according to gender: 'As men and women reflect on the miseries of exilic life, their laments reveal profoundly different senses of what it might mean to inhabit the geographic and social margins of the world' (115). During the years in which the Grendelkin harass Hrothgar's people, Grendel alone crosses the boundary into Heorot: only after he has been killed does Grendel's mother follow. Unlike Grendel, however, who stays in the hall to kill and eat his prey, Grendel's mother grabs one man and flees. Moreover, unlike the female exiles of Klein's analysis, Grendel's mother has power, not only over the place she inhabits but also over the other creatures that live there. She is opposed to Beowulf and to Grendel, but at the same time to normative femininity, thus destabilizing those oppositions. The dwelling she occupies appears to have been a naturally occurring cave, in opposition to the built environment of Heorot, the construction of which had been a collaborative effort among 'manigre mægþe geond þisne middangeard' ('many people throughout this middle-earth,' l. 75). Yet the cave is also opposed to the forest around it, for it is a place where water burns and animals refuse to enter.

The instabilities in how to read Grendel's mother's mere are underscored by the fact that the mere also bears resemblance to the Anglo-Saxon idea of hell, as observed by Carleton Brown in 1938 and affirmed by numerous other readers since then. In Ælfric's rendering of hell, St. Julian tells his tormenters that they will sink 'on ðone sweartan grund' ('into that gloomy abyss,' l. 383) filled with 'undeadlice wyrm / þe eowre lichaman cywð' ('a deathless serpent that will gnaw your body,' *Lives of Saints* 112, ll. 385-86). Such affinities, along with the familial relationship with Cain, suggest that Grendel's mother and Grendel are demonic rather than having any part in the natural world: they are unambiguously neither animal nor human, and perhaps neither. For the Anglo-Saxons, 'the sea both separates and connects,' as Alfred Siewers observes of archipelagic cultures (39). The sea, like Grendel's mother and her dwelling, is unstable, as suggested by the Old English 'Storm' riddle: subject to being flung into the sky and tossed into waves that threaten humans, whether living at its side or attempting to travel over its surface. Through both waves and sea-creatures, the sea and Grendel's mere both pose a direct threat to Beowulf as he attempts to travel through them.

The possibility of multiple readings for the Grendelkin – human, animal, monster, or demon – destabilizes dualistic interpretations that assign either

Grendel's mother or Grendel to any single category, whether affiliated with or opposed to humanity. Ecofeminists reject such dualistic categories and the hierarchies that order them. Patrick Murphy calls instead for a mode of thinking he calls (after Bakhtin) 'dialogics' which 'reveals that the most fundamental relationships are not resolvable through dialectical synthesis' (1995: 3). Karla Armbruster warns against the problems ecofeminists face of, on the one hand, identifying women as having a special bond with nature or, on the other hand, emphasizing too strenuously the distinctions between humans and nature, as well as distinctions among different humans based on characteristics such as race or gender: 'The path between continuity and difference that ecofeminist theorists must walk is so narrow and difficult not because of inadequacies in the theorists or the theories, but because of the complexity of their task' (98). The task of engaging effectively with variously intersecting theories is also, comparably, complex. Armbruster suggests the need for a theory of human subjectivity that acknowledges the multiplicity and constantly shifting quality of the positions that individuals occupy in society, and the ways in which they challenge social orders but are also shaped by them, so that 'we can acknowledge the ways that each person's socially constructed subjectivity is different from that of others without inevitably isolating us from each other' (105). The varying subjectivities and the irresolvable problem of the identity of the monsters in *Beowulf* may help contemporary humans to understand and challenge the ways in which our own shifting identities are constructed.

Menstrual Blood and Amniotic Flood: *Andreas*

Whereas in *Beowulf* the feminine becomes monstrous and is pushed to the margins of both social and natural worlds, in *Andreas* blood and water appear at both center and margins as stand-ins for the fully abjected maternal in the absence of any explicit feminine. While 'wif and wer' ('woman and man,' l. 1597) are referenced collectively, all of the characters given any individuation – guards, cannibals, prisoners, Andreas' torturers, along with Matthew and Andrew and Andreas' retinue – are all male. The youths who return from drowning death to become Christians are likewise all male. The idea that women have any place or function in this social world is thoroughly repressed. As P. H. Cullum and Katherine Lewis point out in their 'Introduction' to *Religious Men and Masculine Identity in the Middle Ages*, 'clerical masculinity was generally formed in relation to other masculinities, not in relation to women, and for many clergy women were irrelevant' (4). Women's positions were defined in terms of men, as the presumed norm, rather than the reverse.

For Kristeva, blood is linked inextricably with the maternal, and both are abjected: defined as impure and made taboo. Kristeva notes that in the Hebrew Bible, the taboos connected with women who have recently given birth are discussed in close conjunction with dietary issues involving the avoidance of eating blood. 'Dietary abomination thus has a parallel – unless it be a foundation – in the abomination provoked by the fertilizable or fertile female body (menses, childbirth)' (100). Jewish dietary law specifically prohibits consumption of blood; Christianity reverses this in the ritual consumption of wine and bread symbolizing Jesus' blood and body. In consuming the blood of their victims, the Mermedonian diet becomes a multivalent symbol within the poem, simultaneously an inversion of Jewish law and a profanation of Christian practice.

After Andreas arrives in Mermedonia, the focus on blood shifts from its consumption to its presence as evidence of bodily injury; simultaneously, the location of blood shifts from the social center of the built city to the wilderness surrounding it. As he is dragged through the wilderness, Andreas' blood flows from wounds all over his body: 'Blod yðum weoll, / hatan heolfre' ('blood surged out in waves, hot gore,' ll. 1240b-41a). The wounds heal themselves each night because of the miraculous intervention of the angel who visits Andreas to lift his spirits, but the ebbing and flowing of blood also suggests menstruation, which recurs and ceases at regular intervals through the middle years of a woman's lifetime.

The multivalent blood also suggests childbirth, an association strengthened by the presence of uncanny water that appears later in the poem. The women who would give birth to and nourish the Mermedonians in the poem, all men – guards, executioners, the youths who are reborn at the end of the poem – are absent. Women, from whose bodies blood flows as a regular, 'natural' part of life, are suppressed from the narrative.

Andrew's body is dragged through caves and across rocks as far as the roads can take them:

Drogon deormodne æfter dunscræfum,
ymb stanhleoðo, stærcedferþne,
efne swa wide swa wegas to lagon,
enta ærgeweorc, innan burgum,
stræte stanfage.

They dragged the courageous one, stout of heart, through hill-caves, around rocky cliffs, even as far as their paths extended, old work of giants, streets paved with stone, inside their cities. (ll. 1232-36a)

The fact that Andreas is dragged through caves, along with the reference to the old work of giants, recall the location of Grendel's mother's lair in a cave beneath a marsh, as well as Beowulf's recourse to a sword described as the ancient work of giants, as the only weapon that will kill Grendel's mother and behead Grendel. The uncanny provides a link here between Andreas and Beowulf – as poems, and as heroes. As Alexandra Bolintineanu has noted, there are numerous parallels in the descriptions of the landscapes in *Andreas* and *Beowulf*. As noted above, Lindy Brady persuasively links the landscape of Mermedonia to the English fenlands, borderlands inhabited by Britons and the Grendelkin. The similar references to caves found beyond the limits of human habitation and the references to 'eald enta geweorc' strengthen the association of blood with the absent feminine in *Andreas* in a parallel to the monstrous feminine in *Beowulf*.

The demons return Andreas to his prison, described specifically with reference to frozen water: the night brings 'snaw' ('snow,' l. 1255), 'hægelscurum swylce hrim ond forst' ('hail-showers and also rime and frost,' l. 1257), and 'cylegicelum' ('icicles,' l. 1260). The passage concludes with the detail that 'is brycgade / blæce brimrade' ('ice bridged the dark sea-road,' ll. 1261b-62a). It requires a truly 'wintercealdan niht' ('a cold winter night,' l. 1265a) indeed, for the sea itself to freeze over. Ice can bridge rivers, making it possible to cross over rather than ford by wading through them. Here, the language suggests a sea frozen solid, rendered impassable by ship and thus cut off from other civilizations through normal navigational processes. In Riddle 69, ice is a miracle; here, it is a further instrument of Andreas' isolation from his community. This frozen water cannot kill Andreas any more than can the injuries inflicted upon him by Mermedonians and night-time demons, nor can it affect his soul. In contrast, the rushing water that Andreas will call forth will kill Mermedonians and then lead, for some of them, to a rebirth of both body and soul: bodies returned to life, souls baptized and converted to Christian faith.

The miracle Andreas performs by causing water to gush forth from a stone occurs, like his torture and bleeding and like Beowulf's slaughter of Grendel's mother, at the margins of the social world:

> He be wealle geseah wundrum fæste
> under sælwage sweras unlytle,
> stapulas standan, storme bedrifene,
> eald enta geweorc ...
>
> He saw by the wall, wondrously fixed under the castle wall, pillars, not small, columns standing, storm-beaten, old work of giants. (ll. 1492-95a)

In an echo of the perverse feast of blood that opens the poem, the water that Andreas sends forth from the ground is characterized as 'biter beorþegu' ('bitter beer-drinking,' l. 1533), an excessive feast of beer and ale that overwhelms the people of Mermedonia. Water is not just water, blood is not just blood; both carry metaphorical and symbolic freight. Riddle 84, conventionally solved as 'water,' identifies its solution also as 'mother.' Riddle 33, solved as 'ice' or 'iceberg,' also uses feminine pronouns to refer to water and to ice, and calls them monstrous and fearsome (see Chapter 6 for a fuller discussion of these two riddles). In *Andreas*, monstrosity is masculine. However, the feminine associations of monstrosity in Anglo-Saxon culture as expressed in the examples Grendel's mother and the Ice riddle, combined with the uncanny freezing of the sea in the cold night of Andreas's captivity, suggest the repressed feminine in the poem.

In Anglo-Saxon biblical lore, the earth is Adam's womb (Estes 643). The water that gushes forth from the earth of Mermedonia spews forth literally from an ancient wall, the work of giants, and metaphorically from the earthly womb. Much as, in *Beowulf*, Grendel's mother's lair functions simultaneously as a tomb for Grendel and as a womb from which Beowulf is reborn into a new social status, the flooding earth of Andreas into which the Mermedonians are swept provides for some a tomb, an 'eorðscræf egeslic' ('horrible cavern/sepulcher,' l. 1588), and for others a womb-like rebirth when they are swept back out of the depths by a reversal of the previously deadly, now life-giving, amniotic water.

While blood is abundantly present in *Andreas*, the maternal and indeed the feminine is entirely absent, suppressed and/or repressed, yet the feminine and the maternal reappear in the water that gushes forth from the earth at the margin of the Mermedonian city, analogous to the marshy borderland lair of Grendel's mother in its simultaneity of death and (re)birth, its evocation of gushing amniotic fluid and its transformation into a baptismal flood as the young people, their youth emphasized and repeated, are brought back to life and simultaneously converted to Christianity. The idea that baptism is a kind of re-birth is not stated explicitly in *Andreas*, but the metaphorical association was known in Anglo-Saxon England: the terms 'eftboren' and 'eftacenned' (both meaning 'born again') gloss Latin 'renatus' ('reborn'), with specific reference to baptism in the late ninth/early tenth century *Durham Ritual* as well as with reference to water and the Holy Spirit in the tenth-century *Rushworth Gospel*.

Conclusion

Water, earth, femininity, monstrosity and death are linked, implicitly and explicitly, across *Exodus*, *Andreas*, *Elene*, and *Beowulf.* While clerical masculinity is constructed without reference to women, the masculinities of Beowulf, Andreas, and the Hebrew warriors of *Exodus* are established against the assumed, essentializing feminine monstrosity of earth and water. As Andreas strides through the flood, it disappears wherever he walks, in an echo of the Red Sea crossing, where God separates the water so that the Israelites can cross and then causes it to flood back, drowning the pursuing Egyptians. The process in *Andreas* is reversed, with fiery water crashing down on Mermedonians followed by its retreat and their baptismal and literal rebirth. Reading these poems from an ecofeminist perspective, which foregrounds both gender relationships and attention to the poems' physical environments, rewards concentration on the figuration of Grendel's mother, as well as the space in which she lives, and on the figuration of water and earth as symbolic of the abjected feminine and maternal. Grendel's mother and her swampy home, the flooding waters and seas of *Exodus* and *Andreas*, and the flowing blood in *Andreas* pose challenges to culturally entrenched dualities between male and female, human and non-human, and land and water. The water that flows out of an age-old stone in *Andreas* and the stone-like sea-ramparts that crash in waves over the Egyptians likewise pose challenges to culturally entrenched dualities between human and nature, because these waters are called forth by humans with the assistance of God, but are not human, and while emanating from or returning to the earth, do not behave in ways that are 'natural.' They therefore suggest an implicit challenge to the distinction between human and nature as well as between male and female, insisting on those boundaries to the point of fracture. Reading these texts ecocritically can also assist in re-thinking floods brought on by climate change today, floods that have human causes at their root and are thus not precisely 'natural,' yet also not precisely 'human.' Insisting on a sharp delineation between masculine and feminine enables and coincides with the attempt to draw a clear line between human and nature; yet, as the Old English poems discussed in this chapter demonstrate, such lines are fuzzy at best, always subject to question and challenge.

The chapter that follows moves from seascapes to ruins. *Andreas* demonstrates the potential for overlap, as the flooding seas literally ruin the city of Mermedonia as well as the bodies of many of its people. The Old English *Exodus*, as noted in brief here, uses figurations of ruined fortresses in its narrative of sea-crossing; these are examined in greater detail. I continue and extend my reading of *Beowulf*, discussing Heorot and the dragon's lair as (future and past) ruins. Analogously, the Tower of Babel in *Genesis A* is depicted as a future ruin even as its construction is narrated, in a scene expanded dramatically from the biblical text. Latin literary inheritances exist alongside remnants of England's Roman occupation, as described in *The Ruin*.

The Old English adaptation of *Exodus* likewise represses the feminine, eliminating references to the biblical Tziporah and Miriam. Similarly, the *Guthlac* narratives include no reference to women, except as objects of rape. *Exodus* is examined in greater detail in the next chapter, on ruins, and *Guthlac* in chapter 4 on wilderness, though neither chapter addresses gender issues. Chapter 6, the second of two on the *Exeter Book* riddles, returns to a discussion of how gender and environmental issues are embedded with one another in Anglo-Saxon literary remains, using Timothy Morton's 'hyberobject' as a way of thinking about cultural formulations like gender, religion, and race.

* This chapter first appeared as "*Beowulf* and the Sea: An Ecofeminist Reading," in *The Maritime World of the Anglo-Saxons*, ed. Bill Schipper, Stacy Klein, and Shannon Lewis-Simpson (Essays in Anglo-Saxon Studies 5). Tempe: Arizona Center for Medieval and Renaissance Studies, 2015: 209-216.

3 Ruined Landscapes

Introduction

Contemporary ecocriticism sees landscape, and our ecosystems, in constant flux, in contrast to earlier literary analysis, which typically understood the places across which textual characters move as 'setting.' Setting is understood to constitute an unchanging background for the actions and emotions of the characters. If setting changes, non-ecocritical literary analysis understands this as occurring in response to the actions of a human 'character' or in metaphorical reflection of a character's spiritual or emotional state. Locations described in several Old English poems – Heorot and the dragon's barrow in *Beowulf*, the Tower of Babel in *Genesis A*, and the crumbling structure(s) of *The Ruin* – are anything but static, and this mutability is in some cases described in terms of non-human forces. The sea in the Old English *Exodus*, discussed with these other ruins because it is transformed in the poem into a stone fortification, is another very changeable environment. All of these texts concern people and events far in England's past and, other than *The Ruin*, also geographically remote.

Yet they also reflect the physical reality that Anglo-Saxon England contained a large number of ruins, unmaintained structures left behind by the Romans. Christine Fell writes that, during the Anglo-Saxon period, 'there must have been a fantastic number of ghost-towns or ghost-villas' surviving from the Roman occupation (179-80). Much as the occupants of England who preceded the Anglo-Saxons left behind traces in earth and stonework, the people who traveled from Scandinavia to become the Anglo-Saxons retained literary and linguistic traces of their pagan past(s) in their names for places, epic heroes, and days of the week. The nature of individual dwellings and community structures gives significant but indirect insight into relationships with the natural (non-human) world. Anglo-Saxon descriptions of their own buildings, and particularly their focus on ruins, gives us a view into a culture that sees itself as built on the remains, literary and physical, metaphorical and literal, of previous cultures. The description of God's interventions in the poetic adaptations of the Tower of Babel episode in *Genesis A* and the Red Sea crossing in *Exodus* contrast with depictions of naturally occurring decay and human destruction in *Beowulf* and *The Ruin*. Taken together, these works suggest that the Anglo-Saxons had a sense of place as fluid and mutable, whether because of divine, human, or natural forces, rather than as fixed and stable.

Nicholas Howe demonstrated that the Anglo-Saxons understood the Exodus passage across the Red Sea as a precursor to their own migration to England. The layering of Jewish, Christian, and English interpretations of Hebrew Scripture, and the complex interactions of Geats, Danes, and Grendel as occupants of the hall, suggest a hybrid sense of place.

In a survey of recent ecocritical work on place in American literature, Karen Halttunen calls for an understanding of place that is 'open ... hybrid ... progressive ... dynamic' (2). The sense in Old English poems of dwelling as impermanent, threatened by human as well as inhuman forces, interacts interestingly with contemporary ecocritical theories of place shaped by global commerce and migration. Julia Hell and Andreas Schönle point out that 'ruin can embody both continuity and rupture, thus inspiring contradictory ideologies and programs.... Narratives of ruination imply human agency' (9-10). Old English poems about ruins suggest different kinds of human agency in the construction of an edifice or in the destruction of the society that built it. But they also describe ruins as the result of natural forces causing decay over time, as well as of divine intervention. The Anglo-Saxon descriptions of ruin in *Genesis A*, *Exodus*, *Beowulf*, and *The Ruin* suggest that the writers and scribes had already developed an understanding of environmental change across both short- and longer-term time frames. The depictions of decay as a result of the actions of nature and time suggest a sense of the earth as a living organism, not static, subject to its own processes outside of human activities or concerns.

The Anglo-Saxons use the idea of ruin as a way of commenting on their own relationships with dwelling, with landscape, with questions of territorial alienation and belonging. Hell and Schönle write that ruins exist only if there is someone to see them: 'The beholder defines the ruin, and the ruin could not exist without such creative appropriation. As a result, the ruin is often the playground of speculative strategies that tell us more about the beholder than about the ruin or its original environment' (7). The poems discussed in this chapter concern themselves with pasts that the Anglo-Saxons have adopted as parts of their own history: they consider themselves heirs to Hebrew, Christian, and Roman traditions. *Genesis A* and *Exodus* appropriate (imagined) Jewish and Hebrew/Israelite pasts; *Beowulf* recalls the Anglo-Saxon tribal past in Denmark; *The Ruin* evokes the Roman occupants of England prior to the Anglo-Saxon settlement. As Helmut Puff observes about post-war Germany, a discussion of ruins is 'a discourse that ultimately is not about the material remains themselves but about self-assurance and alienation.... The fact that ruins are versatile matter, discursively speaking, is one of their most enduring features' (253). Similarly,

Anglo-Saxon poems about ruins from distant pasts and/or remote regions concern themselves with the Anglo-Saxons' own anxieties about loss and displacement. The existence of ruins, whether visible in stone or imagined in poems, challenges how the Anglo-Saxons think about themselves and their relationships to dwellings and to other features of their lived environments.

Hell and Schönle argue that 'catastrophe, real and imagined, underpins modernity's multilayered sense of history' (9). As with many claims about modernity, this too is anticipated within the partially ruined poetic remnants of the Anglo-Saxon period. Poems about the abandoned cities of Babel and Bath, the imagined ruins of stone walls within the crashing waves of the Red Sea, and the burning of a hall like Heorot demonstrate a complex and multi-layered sense of history: the Anglo-Saxons remember their own past in Denmark and interpret it in terms of the Israelite and Roman pasts. They also display a sense of human dwelling as embodied within the natural world, both human and natural intertwined as part of a living organism subject to its own processes beyond human activities or concerns that include natural environmental dynamics across the slow creep of time, or acts of God. Various chronological layers interact with literal and symbolic understandings of time and its meanings.

Roman Past and Mutable Present

The Romans had built cities and fortifications around England, including London, Colchester, York, Gloucester, Lincoln, Canterbury, Worcester, and Exeter (Palliser 18-20), but these fell into ruin when their armies and administrators withdrew. This resulted from a decrease in population leading to the abandonment of buildings as well as from a shift in construction materials and technologies: Romans built with tile and stone, Anglo-Saxons with timber. Early Anglo-Saxons settled near the Roman city of London, rather than within its ruin-filled Roman walls. During much of the period, Anglo-Saxon settlements remained small and non-urban in character, with political and economic activity distributed in various regions rather than centralized in one or a few major cities. By the time of the Domesday Book, there were perhaps three dozen towns in England with a population greater than 1000 (Holt passim). London became a center of trade and commerce in the eighth century, reaching a peak population of somewhere between five and ten thousand inhabitants (Keene 188). Only in the tenth century did the city shift back to its Roman location, occupying the walled Roman enclosure, with streets linking several market areas within the town.

As the Anglo-Saxon period was coming to a close, London began a period of rapid expansion in size alongside the development of its political and economic importance. This continued under the Normans, so that in the late Middle Ages 'London was certainly a major city of the Latin West' (Keene 195). But that takes us well beyond our period. Rome, in contrast, had been a city of a million at its imperial peak and had shrunk to a fragment of its former size, but with an estimated population of somewhere between thirty and fifty thousand at the end of the first millennium, it was still several times the size of London. Moreover, it was in many ways central to European political, religious, and economic life; in contrast, England tended to avoid centralization of various functions in any single location so that its towns did not become true urban centers until the later Middle Ages.

The Old English poem known as *The Ruin* survives in only one manuscript copy, itself badly damaged in what Donoghue calls 'an uncanny instance of form mimicking content' (47). Much of the poem, in fact, cannot be reconstructed, but it is clear that it concerns a ruined stone building or perhaps even a city containing baths, which are mentioned more than once in what survives of the poem. Scholars have suggested that the poem refers to the ruins of Bath, but London or another city is also possible – or, indeed, an imagined conflation of various Roman ruins still visible in many places in Anglo-Saxon England.

The Ruin is first described as the work of giants, broken by fate. This is followed by a description of time and storm and freezing weather damaging the mortar between the stones, according to the dictates of fate:

> Wrætlic is þes wealstan, wyrde gebræcon;
> burgstede burston, brosnað enta geweorc.
> Hrofas sind gehrorene, hreorge torras,
> hrungeat berofen, hrim on lime,
> scearde scurbeorge scorene, gedrorene,
> ældo undereotone.
>
> This building-stone is wondrously broken by fate, the city burst, the work of giants corrupted. The roofs have fallen in, the towers are ruined, gates destroyed, the mortar jutting with frost and the storm-shelter falling down in shards, undermined by old age. (ll. 1-6a)

Internal rhyme and end-rhyme emphasize the inevitable forces of decay. Fate again is blamed for the fall and destruction of masonry and iron, bright halls and baths: 'oþþæt þæt onwende wyrd seo swiþe' ('until mighty fate changed that,' l. 24). The deaths of the inhabitants and then of those able to repair the building are then told: 'betend crungon,' those who would re-build (or those who should have made amends) perished (l. 28), causing 'hryre' (l. 31) – destruction, or decay, or a fall from a height: all of these meanings are possible, so the lack of people to repair the structure could mean that it is left to freezing and storm or to animals digging, burrowing, or leaving herd-paths, or vegetation growing in and around it and causing further decay by its roots, or stones falling from its heights, damaging the lower stories as roofs and upper walls collapse. The poem refers specifically to the absence of the human in the ruins, leaving the natural world as the active force in their decay and destruction. The natural world is changeable, not static, and not even as a background for human activity: as traditionally conceived 'setting,' the ruin is all setting, no human actions.

Russell Berman comments on the ruinous capacities of nature in a North American context: 'the fire, the tornado, the tsunami all devastate buildings, cities, and societies.... Among the weapons of nature, time deserves a special place, since no distinct catastrophe takes place... the silent and slow demolition of age produces destruction as time passes inexorably' (105). Time alone does not cause ruin: flowing water, the cycle of freeze and thaw, plant growth, bacterial action, and other natural processes crumble mortar and cause stone to fall. Berman contrasts the speedy ruination of storm with the slow damage caused by raindrops, repeated cycles of freeze and thaw, the slow growth of plants or the slow action of tides on stone. 'Weapon' is an interesting choice of words for natural destruction, whether sudden results of storm or fire, or the slow decay of time. It refers to a tool used by a being with agency and intention. In the 'Introduction' to *Prismatic Ecology*, Jeffrey Jerome Cohen ascribes an artist-like intentionality to the curves traced by the Mississippi River through North America, calling the natural processes and flows that form the riverbed a form of art: 'The powerful river exerts a relentless agency easily readable in its engendered worlds' (2013: xix). While several of the other poems in the Old English corpus ascribe ruin specifically to divine or human forces, as discussed in greater detail below, *The Ruin* makes no mention of God and explicitly states that decay occurs because of the absence of the human. The notion of nature as an agent of destruction resonates with the description of the storms in the *Exeter Book* riddles, in which nature is pitted not only against human constructions but also against nature itself, churning up waves and felling trees.

Berman argues that 'Ruin is a result of culture, not of nature. Not only does the term "ruin" indicate the destruction of prior human construction, it also suggests human agency. Even where we think we might attribute the demolition work to the unbridled elements or the ravages of time, we soon have to concede that cosmic powers won dominion over the ruined body only because no protective hand was available to guard it' (106). *The Ruin* makes a similar point: the stone structures have fallen because those who could repair them have died. Berman's metaphorical use of the word 'unbridled' evokes a domesticated horse that is not, at a particular moment, under human control, in turn suggesting that nature could, at least in some instances, be controlled by humans; his use of 'ruined body' to refer to the destruction of something previously built by humans is also suggestive. Berman's discussion of ruins is not concerned with environmental or ecocritical issues, focusing instead on human relationships to ruins. However, his work is valuable in terms of thinking about these texts ecocritically because of his insights about different kinds of ruination and different time-scales. Human attention cannot always keep built environments from ruin: catastrophic destruction, whether from natural causes such as storm or flood, or human causes such as war, cannot always be staved off; and wooden beams decay over time in spite of attention.

Hell and Schönle point out that different time-frames are brought into play whenever someone sees or reads ruins. 'Ruin gazing ... always involves reflections about history: about the nature of the event, the meaning of the past for the present, the nature of history itself as eternal cycle, progress, apocalypse, or murderous dialectic process' (1). They then leap, as so much contemporary theoretical scholarship does, from ancient to modern, Rome to Romanticism, in their consideration of how ruins have functioned and what they have meant to different cultures. Ignoring the medieval period altogether enables their claim that 'only in a secularized world do ruins become objects deemed suitable for study or interpretation... . What we now call ruins began to be perceived and preserved as such during the Renaissance' (5).

Perception of ruins, and their study and interpretation, are distinct from preservation. Beasley-Murray argues that 'Ruins are presented as foundation,' and points out that 'sometimes this foundation is literal, as when the stones of earlier constructions are built over or recycled for new buildings. But this recycling threatens to undercut the linearity of the history that it otherwise anchors: if one civilization can fall to ruin, then so perhaps can – or must – each subsequent one' (214). Roy Liuzza argues that for Bede, ruins were important as 'intelligible evidence' in support of his history, signifying continuity rather than cultural rupture (8). Reuse of materials from Roman ruins in Anglo-Saxon and later medieval English structures, including churches, walls, bridges, secular buildings and constructions, is well-established. An interesting example is the reuse of a Roman sarcophagus from Grantchester, brought to Ely for use as a tomb for the Anglo-Saxon abbess Æthelthryth. This is a different kind of preservation, one that inscribes onto a single object several layers of meaning, preserving its form and, in this case, its function while reusing it in a different temporal and religious context. Though Anglo-Saxon builders did not preserve the physical remains of *The Ruin*, at least in part because they lacked the skill to build or repair stone, the poem preserves what it perceives in textual fashion. The Anglo-Saxons did not preserve ruins in the way that Hell and Schönle construct as appropriate, but they clearly saw them, and they recycled and reused them in their own projects, literary as well as architectural.

Imagined Biblical Origins

Given the prominence of ruins in the historic Anglo-Saxon landscape, perhaps it is not as remarkable as it seems at first that ruined buildings are added to the poetic versions of both *Genesis A* and *Exodus*. Liuzza calls the Tower of Babel episode in the Old English poem a 'remarkably expansive re-imagining' (4) of the biblical narrative, with added detail about the physical structure of the city and tower. In *Genesis A*, God is responsible for dispersing the people so that construction is interrupted and the building, left incomplete, will fall rapidly into ruin, in an analogue to the stone buildings left behind by departed Romans in *The Ruin*, which fall into disrepair because the people who would maintain them are absent. In *Exodus*, the flooding Red Sea is described not only as drowning the Egyptian warriors and sweeping their chariots away, but also as destroying stone walls and ramparts. Liuzza argues that Old English poems about ruins draw on Latin exegesis in addition to original sources, but 'while Latin poetic elegies tend to use fallen cities as warnings of the passing of all earthly glory, patristic and medieval exegesis of the story of Babel was the site of a confluence of discourses on language, history, and pride' (6). Both poems include ruins in places of departure, following the biblical text for Genesis but added in the poetic *Exodus* without parallel in the biblical book. All of humanity is dispersed across the world from Shinar after God halts the construction of the Tower; the Egyptians are swept away in defeat and the Israelites travel beyond the Red Sea in the *Exodus* narrative. The physical landscape of Anglo-Saxon England contained ruins in the places the Romans had left behind; the absence of the people who had inhabited them leads to their decay and eventual destruction.

The expansion of the Tower of Babel episode from the biblical version into *Genesis A* includes additional details about the construction of the tower and its incomplete state, with six brief biblical verses expanded into nearly 30 lines of poetry. Beasley-Murray suggests that 'ruins are incessantly seen as pointing beyond themselves, to some absent totality' (215). The unfinished Tower in Shinar points to an essentially mythological explanation for the dispersal of the human population across the globe and for the differences in language by way of asserting the power of God over humans. In the Vulgate text, the description of the tower is brief:

> Dixitque alter ad proximum suum: Venite, faciamus lateres, et coquamus eos igni. Habueruntque lateres pro saxis, et bitumen pro cæmento: et dixerunt: Venite, faciamus nobis civitatem et turrim, cujus culmen pertingat ad cælum.
> And they said to one another, come, let us make bricks, and bake them in fire. And they had bricks for stone, and pitch for mortar, and they said, Come let us make ourselves a city and a tower, whose peak extends to heaven. (Gen. 11: 3-4)

God descended 'ut videret civitatem et turrim' ('and saw the city and the tower,' Gen. 11: 5). The final reference to the structure comes at the end of the passage, as the people 'cessaverunt ædificare civitatem' ('stopped building the city,' Gen. 11: 8). After the initial reference to bricks and mortar, there is no further mention of building materials or of the status of the construction. In the Old English poetic adaptation, in contrast, the three verses describing the construction and its cessation are expanded significantly. The *Genesis A* version extends the Vulgate description of the people planning the structure and comments more fully on their pride and sin.

> laras sohte
> weras to weorce and to wrohtscipe,
> oðþæt for wlence and for wonhygdum
> cyðdon cræft heora, ceastre worhton
> They sought out teaching, men for work and for crime, until out of pride and recklessness they made their craft known, built a city. (ll. 1671b-74)

The Old English *Genesis* also expands the description of the structure itself, calling it 'burh' and 'beacne torr' ('city' and 'tower as beacon,' l. 1666), as well as 'ceastre' ('city,' l. 1674) and 'steannene weall' ('stone wall,' l. 1676). As in the Vulgate and the Old English prose versions, God comes to see the work the men have done, scrambles their language, and scatters the people across the earth. The *Genesis A* account then repeats details about what the builders have attempted, referring once again to the 'weall stænenne' ('wall of stone,' l. 1691), but using the verb 'timbran' ('build with wood,' l. 1692) to refer to the construction of the stone wall, reflecting Anglo-Saxon building practices and lack of knowledge about working with stone. The Vulgate provides no further detail about the construction, stating simply: 'divisit eos Dominus ex illo loco in universas terra et cessaverunt aedificare civitatem' ('God divided them from their place into diverse lands and they stopped building the city,' Gen. 11: 8). The Old English adaptation, however, describes the departure of the people and then references once again the building that has been left abandoned: 'Him on laste bu / stiðlic stantorr and seo steape burh / samod samworht on Sennar stod' ('In their tracks both the strong stone tower and the high city stood together, half-finished, in Sennar,' ll. 1699b-1701). The 'civitatem' ('city') is reimagined as both 'burh' ('city') and 'stantorr' ('stone tower'). The isolation of the structure, unpeopled, recalls the separation in *The Ruin* between stone and dwellers, the buildings deprived of people who could repair them. As noted above, the Anglo-Saxons had themselves migrated from one place to another, and they lived in a place where other people had preceded them, and then been scattered, as recognized in *The Ruin* as well as in the *Anglo-Saxon Chronicle*'s record of the Roman occupation of England. The narrative of linguistic dispersal might have had special relevance for a people living in a place in which two cultures and two languages jostled for space. I have argued elsewhere (Estes 2007) that *Genesis A* might have been part of the late ninth-century Alfredian program of translation, undertaken after the establishment of the Danelaw. The emphasis in this passage on the construction of the tower, abandoned because the people could no longer understand one another, may be another piece of evidence for the production of this poem in or shortly after a time when the English had been negotiating with Danes for territory and confronting the need to conduct trade and other transactions in a time of linguistic difference and instability.

The monastic population that would have been responsible for the adaptation of biblical materials into Old English heroic verse, and its subsequent manuscript inscription, took what Old Testament scholars call a 'Christological' interpretation of the material. As Rosemary Radford Ruether and James Parkes have pointed out, readings of Old Testament narratives undertaken from a Christian perspective read them not as independent attestations of Jewish religious development but as anticipating the split between Christianity and Judaism. The figures in the text, then, are read as anticipating either Christians (positive) or Jews (negative), with occasional use in Old English of 'Hebrews' as an apparently neutral term. Thus Abraham, Isaac, Jacob, Daniel and a few others are interpreted as proto-Christian heroes anticipating the adoption of Christian faith, while most of the Hebrews and Israelites are seen as anticipating the 'Jews' whom many people, clerical and lay alike, believed had killed Jesus. A poetic adaptation of the Tower of Babel episode done in a Christian context needs to be understood within this framework. The Tower of Babel narrative is taken in Christological exegesis to demonstrate that even after the killing of almost all of humanity in the Flood, humans are, with rare exceptions, degenerate, requiring redemption through the eventual arrival of Jesus. In *De Initio Creaturae*, Ælfric follows the account of the dispersal of languages and peoples immediately with an account of all of the people worshiping idols – with the one exception of Noah's son Shem, grandfather of Eber: 'of þam asprang þæt hebræisce folc. þe god lufode. & of þam cynne comon ealle heahfæderas. & witegan þa ðe cyddon cristes tocyme' ('Of him originated the Hebrew people, who loved god, and of that people came all of the patriarchs and wise men, those who proclaimed the coming of Christ,' 118: ll. 227-29).

The focus in the episode on the physical ruin of the structure of the Tower is multiply meaningful from a Christian Anglo-Saxon point of view. It links the superseded Jews to the Anglo-Saxons' pre-Christian ancestors in Denmark, as well as to the Roman builders of the many stone ruins found in the Anglo-Saxon landscape. Environmental details in *The Ruin* are presented in a landscape devoid of human habitation; the natural world itself had been in a state of unceasing change causing the stone towers to decay over time. The Tower of Babel episode in *Genesis A* locates environmental issues in the context of the relationship between God and humans, and suggests that excessive human labor is subject to divine censure.

The narrative of the attempted construction of Babel followed by the global dispersal of all of the builders, leaving the city abandoned, anticipates or describes environmental degradation, particularly in the Old English version of the narrative, which foregrounds walls and stone and construction. The episode could be read as a myth of origins about the environmental dangers of human excess, but there is no evidence in either the Old English prose translation of the text, which is a nearly word-for-word rendition of the Latin, in Ælfric's commentary on the passage, or in *Genesis A* that an environmentally focused interpretation had any resonance with contemporary audiences. It demonstrates, instead, human excess of pride, which requires correction by God.

The emphasis on stone echoes the crumbling stone walls in *The Ruin*, but the instantaneous conversion of the Tower of Babel from construction site to half-finished ruin stands in stark contrast to the attention in *The Ruin* on natural forces of destruction. As Levi R. Bryant points out, popular ecological narratives assume that without humans, nature will 'maintain sustainable equilibrium' (290). Yet in both *The Ruin* and the Babel narrative, the absence of humans results in environmental transformation. It seems clear from both of these very different narratives that the Anglo-Saxons viewed the natural world as a place of fluctuation and on-going variability, not one of stasis or equilibrium, in cases when humans are absent as well as in consequence of human actions or divine intervention.

The Old English adaptation of the Flood narrative later on in *Genesis A* likewise modifies the Latin narrative to depict the landscape in a state of flux rather than stasis, an agent alongside humans rather than an unchanging setting for human actors. In the biblical text and its Old English analogues, the ark floats for about a year, after which vegetation and agriculture, including sufficient animal feed, were apparently divinely restored. There is no suggestion that buildings or fields have been damaged by the flood, or if they were, they were restored by divine fiat immediately after the floodwaters receded. The only reference to agricultural labor in the biblical text has Noah establishing a vineyard: 'Coepitque Noe vir agricola exercere terram, et plantavit vineam' ('And Noah, a man of agriculture, began to cultivate the ground, and planted vines in a vineyard,' Gen. 9: 20). This passage is rendered nearly word-for-word in the Old English prose translation of Genesis but expanded substantially in *Genesis A*:

ðа Noe ongan niwan stefne
mid hleomagum ham staðelian
and to eorðan him ætes tilian;
won and worhte, wingeard sette,
seow sæda fela, sohte georne
þa him wlitebeorhte wæstmas brohte,
geartorhte gife, grene folde

Then Noah began another time with his kinsmen to build a home and to till the earth for food for himself. He labored and he worked, planted a vineyard, sowed many seeds, and proceeded eagerly that the radiant brightness produced a crop, perennially bright gift, the green field. (ll. 1555-61)

The Anglo-Saxon translator seems to have been as baffled as the modern reader at the ease with which the biblical text assumes the resumption of agricultural labor, and expands to comment that Noah had to begin again to work. The implication in the Old English poem is that the fields previously cleared, tilled and planted had been destroyed by flood and needed to be worked anew. The biblical text suggests that agriculture resumes after a year of flooding by an act of God. The Old English version, conversely, implies an environment not in a state of equilibrium or static continuity, but one damaged by flood and requiring human intervention before it once again can generate produce. As Bryant writes 'Far from having stable and impenetrable boundaries, things are modified as a result of their encounters with other things' (292). The biblical text gives no impression of such modification, instead giving God the power to change the environment instantaneously from one fixed state, flooded, to a different one, unflooded, as if no flood had occurred. The changes made to *Genesis A*, however, suggest an Anglo-Saxon environmental conception in which things affect, and matter to, other things. The Old English text represents agricultural landscape as potentially unstable, an entity in flux and not simply subject to either divine or human interventions, in a more subtle and complex depiction than that of the biblical text.

In *Exodus*, the floodwaters of the Red Sea that kill the Egyptian warriors pursuing the fleeing Israelites are re-imagined as ruined walls and ramparts. The Old English poetic adaptation expands the two biblical verses describing the drowning of the Egyptians into nearly seventy lines of verse, an elaboration even more extensive than the adaptation of the Tower of Babel episode in *Genesis A*. As with *Genesis A*, divine intervention enables the flight of the Israelites and the entrapment and death of the Egyptians. The inclusion of a version of the Flood narrative in the poetic *Exodus* establishes the connection, for Anglo-Saxon audiences, between the two events.

The ruins, described as composing the waves of the Red Sea as the Israelites pass and the Egyptians are drowned, have no parallel suggestion in the original, and point to something beyond the text. Moses raises his hand, and the sea parts and forms itself into walls; the Israelites walk through a dry passage, and then Moses extends his hand once again and the water rushes back. In the Vulgate, the drowning of the Egyptians is described rather laconically, with water flooding back onto the sea bed.

> Cumque extendisset Moyses manum contra mare, reversum est primo diluculo ad priorem locum: fugientibusque Ægyptiis occurrerunt aquæ, et involvit eos Dominus in mediis fluctibus. Reversæque sunt aquæ, et

> operuerunt currus et equites cuncti exercitus Pharaonis, qui sequentes ingressi fuerant mare: nec unus quidem superfuit ex eis.
> And when Moses extended his hand toward the sea, it reversed at the first break of day to its previous place, and as the Egyptians were fleeing, the waters rushed to meet them, and God covered them in the midst of the waves. And the waters were reversed, and entirely covered over the chariots and the horses of Pharaoh, who had gone into the sea after them: not one of them survived. (Exod. 14: 27-28)

In the Old English adaptation, the drowning of the Egyptian army is cast in terms of bloody martial defeat, with much repetition of words for blood, wounds, and gore, and God himself wielding an 'alde mece' ('old sword,' l. 495), reminiscent of the old giants' sword with which Beowulf killed Grendel's mother. More remarkable is the use, within the Old English rendition of the passage, of metaphors for fortification within the sea that are destroyed by its force:

> Randbyrig wæron rofene, ...
> Flod famgode, fæge crungon,
> lagu land gefeol, lyft wæs onhrered,
> wicon weallfæsten, wægas burston,
> multon meretorras.
> The fortified walls were broken.... The flood boiled, doomed ones fell in battle, the sea fell over the land, the air was disturbed, fortifying walls gave way, the waves burst and consumed the towering sea-walls. (ll. 464a, 482-85a)

Later in the passage, there are further references to destruction of the 'flodweard' ('sea-wall,' l. 494) and 'weallas' ('walls,' 'ramparts,' l. 572). Finally, 'þa mægenþreatas meredeað geswealh, / spelbodan eac' ('Sea-death swallowed the mighty band, the messengers as well,' ll. 513-14a). Nature is 'denaturalized': the sea (a place that, as discussed in the previous chapter, humans can neither inhabit nor tame) is imagined in terms of human construction and occupation and reimagined as a kind of landscape.

The 'walls' described in the text are metaphorical: walls of water between which the Israelites had walked now collapsing onto the Egyptians. But they are simultaneously walls constructed of timber and stone, being destroyed by the sea, much as actual flooding periodically damaged Anglo-Saxon towns and fields. Two such floods are described in the *Anglo-Saxon Chronicle*. The entry for 1014 states:

> & on þissum geare on Sancte Michaeles mæsse æfen. com þet mycele sæflod geond wide þysne eard. & ærn swa feor up swa næfre ær ne dyde. & adrencte feala tuna. & manecynnes unarimædlic geteall.
> And in this year on the eve of St. Michael's fast, the great sea-flood came widely across this country, and ran farther inland than it ever had before, and drowned many towns and countless people. (*Two Saxon Chronicles* 145)

The entry for 1125 gives considerably greater detail:

> On þes ices geares wearð swa micel fleod ... þ[æt] feola tunes & men weorðan adrencte. & brigges to brokene. & corn & mædwe spilt mid ealle. & hunger & cwealm on men & on erue. & on ealle westme swa micel untime wearð swa hit ne wæs feola gear ær.
> such a great flood ... that many towns and people were drowned, and bridges broken and grain and pasture entirely destroyed. And famine and plague among men and among the herds. And for all the crops there was more misfortune than there was for many years before. (*Two Saxon Chronicles* 255-56)

Both of these historical floods occur after the inscription around 1000 of the Junius manuscript, in which *Genesis A* and *Exodus* are preserved. But knowledge of the destruction of fields and buildings from flooding seems to be behind the remark in *Exodus* that 'randbyrig wæron rofene' ('the fortified walls were broken,' l. 464) as well as behind the recognition, articulated in the adaptation of the Flood narrative in *Genesis A*, that fields that have been flooded must be worked again. Hell and Schönle ask 'is a ruin an object or a process?' (6). In the biblical text, the ruin of Babel is an object. In its reconceptualization in the poem, the ruined Babel is a process, as are the fields and (metaphorical) towers destroyed by flood in both *Genesis A* and *Exodus*, and the decaying buildings of *The Ruin*.

Imagining the sea floor as containing walls and ramparts that are destroyed by the waters flooding back also domesticates it. As discussed in the previous chapter, seascapes are distinct from landscapes. Land may be wilderness, but (with exceptions like the highest reaches of mountain ranges or the hottest desert) it holds the possibility of adaptation for human habitation, domestication of animals, and agricultural use. The sea, however, is uninhabitable for humans. The Grendelkin are monstrous in part because they can live in an underwater cave. In Exodus, as Denis Ferhatović argues, 'through God's help, Moses tames the hostile landscape, even incites it to turn into a more human habitation' (517). The sea-walls of *Exodus*, simultaneously destroyed and destroying, are at once process and stony reality.

The depiction of ruins within the flooding sea of *Exodus* presents the natural world in purely instrumental form, mutable but also subject to divine will, described purely in terms of human needs and processes. As Hell and Schönle remark, introductory to her essay, Kerstin Barndt sees in the contemplation of ruin 'the pleasures of postmodern spectacle' (Hell and Schönle 12), and there is pleasure in *Exodus* in the play of spectacular language for blood and gore, flood and ruin, or in *The Ruin*'s melancholic narration of slow decay. In the depiction of war as a pageant of drowning and bleeding and dying, the sea is rendered an instrument of human imagination and human concerns. The flood of *Genesis A* is an instrument of God, yet the tilling of the earth is a human activity. The parting and flooding of the Red Sea, in contrast, is entirely the work of God. As Barndt argues, about industrial ruins in Germany, 'Ruins are palimpsests that invite us to contemplate a layered temporality' (270). Reading the 'Old Testament' or Hebrew scripture from a 'Christological' point of view centered on the idea that a Christian reading is the only valid one – whether in the Anglo-Saxon period or today – collapses different temporalities. As Howe argued in *Migration and Mythmaking*, in reconstructing the narrative of Exodus the Anglo-Saxons added to the Hebrew and Christian layers additional folds of meaning relating to their own temporal and geographical histories. In addition, Barndt points to 'contradictions between different temporal layers, material remnants, and human experiences' (273). The idea of the sea as a place of meaning for humans, a place of transition from Egypt to Israel and from Denmark to Anglo-Saxon England, sits in contrasting tension with the reality of the sea: a visible surface with depths unknown to humans, useful as a means of transit and source of food but threatening in the fact that it cannot be settled, that humans can never dwell in it.

Constructed Danish Memories

Beowulf was written down in the same culture that produced the Old English adaptations of Genesis and Exodus. Though it does not explicitly draw on or incorporate Christian narratives or theology, it has long been recognized as a work that took its final shape within a Christian context, memorializing a Danish, pagan past while depicting it as irrevocably superseded, temporally, geographically, and spiritually. The poem's complex interplay of human, divine and natural forces in conjunction with monstrosity makes the poem's depictions of ruins, past and present as well as future, particularly interesting from an ecocritical perspective. As discussed in the previous chapter, the sea in *Beowulf* serves human purposes as diverse as providing a space for transport and enabling deeds of slaughter interpreted as contributing to Beowulf's heroic masculinity. The marsh in which Grendel's mother lives is a liminal space, neither earth nor sea, a place of horror, again deployed instrumentally in the text to illuminate human concerns. From an ecocritical point of view, the depiction of ruins in *Beowulf* provides a more nuanced perspective on the environment, demonstrating the potential for human inability to control environmental factors. The non-human world is, when it comes to ruins, not fixed or stable, but mutable, sometimes because of human actions but other times because of natural processes. It can be seen as an agent, not merely something acted upon, though this is muted and nuanced within the poem.

Beowulf as a whole is a narrative recalling a lost and ruined social world – pagan Denmark, dim in the memories of the people who still pass down the stories. The manuscript itself is another ruin, following the Ashburnham House fire, each leaf carefully conserved by librarians and the whole stitched back together so it can once again be read, minus the burned-off edges. Like *The Ruin*, *Beowulf* describes objects and places as the ancient work of giants. The two works are thus linked as codicological ruins as well as in describing human artifacts ravaged by the forces of time.

Andreas Huyssen points out that ruins give time to place: 'In the ruin, history appears spatialized, and built space temporalized' (21). Yi-Fu Tuan also points out that space and time are intimately connected: 'Distance, unlike length, is not a pure spatial concept; it implies time' (119). Sea journeys in Old English poetry are given in distances of travel time, rather than in spatial units: Beowulf swims for seven days with Breca; he and his retinue sail for two days to reach Heorot; it takes him the better part of a day to reach Grendel's mother's lair. Much as Anglo-Saxon and later medieval structures incorporated stone and tile from Roman ruins, and the tomb of Æðelþryþ repurposes a Roman sarcophagus, *Beowulf* makes use of the ruined materials of oral-formulaic Danish poetry to craft a work that makes sense within the tenth-century context in which the manuscript was written. The use of old legends of heroes in the construction of the epic – whether the materials used in the main narrative, or the stories told as 'digressions' within it, or the ruined graves and caves depicted as part of the main narrative – preserves them for later observation in that it 'reads' them as scenes of value for the Anglo-Saxons to understand their tenth-century conditions. As Hell and Schönle argue, 'The ruin is a ruin precisely because it seems to have lost its function or meaning in the present, while retaining a suggestive, unstable semantic potential' (6). The ruins of *Beowulf* and *The Ruin* are textual, paleographical, and physical, and at a poetic remove from the time of preservation.

In *Beowulf*, the hall Heorot's ruin is foreshadowed in the precise moment when its construction is narrated:

> Sele hlifade,
> heah ond horngeap, heaðowylma bad,
> laðan liges; ne wæs hit lenge þa gen
> þæt se ecghete aþumsweorum
> æfter wælniðe wæcnan scolde.
> The hall towered, high and broad, awaited waves of war, the hostile fire; it was then not yet long until the deadly enmity should awaken the sword-hate of father-in-law and son-in-law. (ll. 82b-85)

Drawing on Anglo-Saxon archaeology, Karl Wentersdorf (2007) suggests that Heorot might be understood as having been built over a Roman ruin:

> The building was approached by a stone-paved roadway (*Stræt wæs stanfah* [320]), undoubtedly of Roman origin.... when the poet referred to the floor of Heorot as 'brightly colored' (*fagne flor* [725]), he may well have imagined the *dryhtsele* as having been erected over foundations like those of the many ruined Roman villas in England, palatial in luxury and decorated with tessellated floors, some of which have survived into the third millennium. (411-12)

In the same essay, Wentersdorf further suggests that the detail of Heorot's gilded roof (referenced, as he points out, three times) was drawn from contemporary Latin descriptions of buildings with golden roofs, as no archaeological evidence survives for an Anglo-Saxon building with bronze or gilded roof tiles. This is a literary depiction drawing on Roman textual foundations of a hall built on a Roman stone foundation. No archaeological evidence exists for Roman villas in Denmark: this is an edifice of English, not Danish, imagination.

In a discussion of the re-use of ruins in a Peruvian village, Beasley-Murray comments:

> Ruins are presented as foundation.... Sometimes this foundation is literal, as when the stones of earlier constructions are built over or recycled for

> new buildings. But this recycling threatens to undercut the linearity of the history that it otherwise anchors: if one civilization can fall to ruin, then so perhaps can – or must – each subsequent one. (214)

The presence of such a large number of ruins, comprising individual buildings as well as entire towns and even the city of Londinium – much larger than the later Anglo-Saxon town of London – would have been a constant reminder to the Anglo-Saxons of the possibility of social, political, and physical collapse. The idea that Heorot was built upon Roman foundations, and the fact that its potential ruin in flames is described at the same time as its construction, both reflect an awareness of the mutability of physical and social structures.

Tuan argues that 'Human groups nearly everywhere tend to regard their own homeland as the center of the world' (149). This would suggest that for the Anglo-Saxons, Denmark retains the status of homeland. Yet as Alfred Hiatt argues, the geography of *Beowulf* is blurred and multiplied:

> *Beowulf* is not a geographic text. It is a poem that describes not a single adventure but several; not a single place, but several; relations between not two peoples but between many. In doing so it constructs something that cannot, ultimately, be reduced to a homeland: the space *Beowulf* writes is regional, a periphery without a centre, whose overriding motifs are exile, mixture, loss – and survival. (40).

In one sense, Beowulf's kingship stands at the center of the poem, framed by his journey to Heorot to slay the Grendel-kin and his death in the teeth of the dragon. In another sense, Heorot-to-be-ruined is central, framed by the ruined dwellings of Grendel and the dragon. *Beowulf* possesses multiple possible centers, geographical, textual, and temporal.

While *The Ruin* describes a building already fallen into decay, *Beowulf* narrates the destruction of the great hall Heorot just after describing its construction. After it is built, however, it becomes a ruin of another sort, made useless by Grendel's night time attacks: 'idel stod / husa selest' ('the best of halls stood idle,' ll. 145b-46a). After the hall has stood unoccupied for twelve years, Beowulf arrives; in the ensuing fight, the hall is badly damaged: 'þa wæs wundor micel þæt se winsele / wiðhæfde heaþodeorum, þæt he on hrusan ne feol' ('It was a great wonder that the wine-hall withstood the brave ones, that it did not fall to the earth,' ll. 771-72). The poem turns to the Danes' rejoicing over the death of Grendel and celebration, and concludes by further describing the extent of the damage:

Wæs þæt beorhte bold tobrocen swiðe,
eal inneweard irenbendum fæst,
heorras tohlidene. Hrof ana genæs,
ealles ansund
The bright hall was badly ruined, the inside fastened with iron bands,
hinges burst. The roof alone survived, entirely sound. (ll. 997-1000a)

Before reiterating how badly damaged the hall has been by the fight, the poet has already discussed its repair. The inverted narration of repair followed by a restatement of the extent of the damage echoes the earlier passage in which the hall's destruction by fire as a result of warfare is foreshadowed immediately after the description of its construction. The final image in the earlier passage is that of the burning hall; here the final image is of doors hanging off hinges, with only the roof unscathed.

What distinguishes Heorot from the hall or halls described in *The Ruin* is the presence of people who can make repairs. *The Ruin* says that those who can repair the hall are gone, dead. While Heorot had lain empty for more than a decade as a result of Grendel's nightly depredations, it has people who can and will repair it. Reading the two poems side by side emphasizes the need that buildings have for humans who can maintain them against decay, thus emphasizing the intertwined nature of built environments and the 'natural' world. Thinking of buildings as needing humans for maintenance as much as humans need them for shelter foregrounds the human-object entanglement, which is explored in considerably greater detail in Chapter 6.

Following the destruction of the hall, *Beowulf* turns to ruined bodies rather than ruined buildings, with Grendel's arm swapped for Aschere's head and then for Grendel's head. Grendel's mother cannot be killed with a normal weapon, but only with 'eald sweord eotenisc' ('an old giants' sword,' l. 1558). Even that sword's blade melts away after Beowulf uses it to cut off Grendel's head; he brings the hilt back to Heorot and presents it to Hrothgar after telling him about his success in battle. (I have not attempted a queer ecological reading in this volume, and thus the intriguing point that the very phallic blade (Chance 253) melts after it penetrates Grendel, but not Grendel's mother, is beyond the scope of the work.) The sword is described again as 'enta ærgeweorc' ('the old work of giants,' l. 1679) a phrase closely similar to the description in *The Ruin* of the building as 'enta geweorc,' ('work of giants,' l. 2). There is much description of the swampy territory in which Grendel's lair is located, of the sea-creatures that inhabit it while deer and other wild animals avoid it, but very little of the lair itself. It is first called 'hof' (l. 1507), a word referring to an enclosed dwelling, often of fairly high status, and 'reced' (l. 1571), a hall or palace, and a word also used for Heorot, as well as 'niðsele' (l. 1513), a hall of evil or strife; 'sele' and several compounds are also used for Heorot. Like the dragon's lair later in the poem, Grendel's lair contains a hoard of treasure, but unlike the dragon's lair, little information is given about it. Beowulf takes with him to the surface of the mere only the head of Grendel and the hilt of the sword he had used to decapitate him.

Interestingly, the hilt of the sword is inscribed with a tale of an ancient flood:

> Hroðgar maðelode, hylt sceawode,
> ealde lafe, on ðæm wæs or writen
> fyrngewinnes, syðþan flod ofsloh,
> gifen geotende, giganta cyn
> (frecne geferdon); þæt wæs fremde þeod
> ecean dryhtne; him þæs endelean
> þurh wæteres wylm waldend sealde.
>
> Hrothgar spoke, looked at the hilt, the old remnant, on it was written the beginning of ancient strife, when the flood killed, the flowing sea, the race of giants (they had done bad things); that was a people foreign to God; God gave them their reward through the welling of waters. (ll. 1687-93)

The tale of ruined landscape is the *Genesis* Flood narrative, in which God inundates the entire world, killing everyone and everything except a representative pair of each kind of living creature (including Noah and his wife as the exemplary humans). This is the first ruined landscape.

James Doubleday points out that Grendel is descended from the giants who attempted to build the tower of Babel, left in ruins after God divides the languages of the builders: 'In his arrogance of power, he [Grendel] shows himself a true descendent of the giants and especially the giant Nimrod, the founder of Babylon (Genesis 10:10) and the supposed builder of the Tower of Babel' (Doubleday 8). While the builders of Babel succeeded in their construction until God caused them all to speak different languages, it appears Grendel cannot build, but only destroy. While Doubleday argues that his dwelling-place in the mere is no mere lair, but a hall to rival Heorot, there is no evidence that Grendel (or his mother) constructed the hall; if it is not a natural cave, it is perhaps a burial-mound into which they have moved, like that occupied by the dragon later in the poem.

The dragon's lair described later in *Beowulf* is an ancient stone burial chamber and treasure-hoard in which the dragon has taken up residence.

The hoard directly echoes that found in Grendel's lair; the underwater cave differs from the dragon's barrow in that it appears to have been a naturally occurring hollow rather than an ancient human construction. Beowulf has become king of the Geats and has ruled in peace for many years before the dragon begins to harry the kingdom:

> oððæt an ongan
> deorcum nihtum draca ricsian,
> se ðe on heaum hofe hord beweotode,
> stanbeorh steapne; stig under læg,
> eldum uncuð.
>
> until a solitary one began in the dark night to do violence, a dragon, he who, in the high hall on the steep rocky hill guarded a hoard; under it lay a path, unknown to men. (ll. 2210b-14a)

The word 'hof' is used in this passage, as for Grendel's lair: once again, a term that can mean a dwelling of high status is used to describe the dwelling of a monster. It calls to mind the references to Grendel's mother as 'ides,' a term in this context of significant ambiguity and perhaps ironic force, as it means elsewhere a woman of high, even noble, status. A few lines later the dragon's lair is specifically identified as a burial chamber ('beorh,' l. 2241) the contents of which had been secured by some sort of imprisoning or artful power ('nearocræftum fæst,' l. 2243). But by the time the dragon encountered it, the art or the craft that had kept the chamber locked had failed: 'Hordwynne fond / eald uhtsceaða opene standan, / se ðe byrnende biorgas seceð' ('An old night-robber found that hoard-joy standing open, the one who, burning, seeks out graves,' ll. 2271b-72). The barrow is another physical ruin, its doors and its physical or magical locks broken or simply decayed, allowing the dragon to enter and to take possession of the hoard of treasure that has been left inside. Unlike *The Ruin*, *Beowulf* does not include a description of the processes that have contributed to the degradation of the barrow, whether human, natural, or supernatural, though the reference to the passage of many year suggests the slow decay of nature. As discussed in depth in the following chapter, Guthlac's barrow – an abandoned grave at least in the prose texts, though perhaps a mountain in the verse adaptation, possesses some similarities to the dragon's lair: a former grave, it has been opened by grave robbers and then used as the site for a cistern, also recalling the watery context of Grendel's lair.

After Beowulf and Wiglaf have slain the dragon, Beowulf sits down beside the barrow and looks at it and 'seah on enta geweorc, / hu ða stanbogan stapulum fæste / ece eorðreced innan healde' ('looked at the work of giants, how the stone arches were forever held fast on columns inside the earthen hall,' ll. 2717b-19). The arched stone and the columns suggest a Roman structure rather than a Saxon one, the detail of columns and arches recalling those of the *Ruin*.

The dragon discovers that his hoard has been disturbed and a cup taken, but it waits for nightfall before emerging to exact revenge. Like Grendel and Grendel's mother, he attacks under cover of darkness, perhaps a strategic tactic, perhaps a marker of non-human status, similar to animals that see best in the dark:

ða se gæst ongan gledum spiwan,
beorht hofu bærnan ...
Hæfde landwara lige befangen,
bæle ond bronde, ...
þa wæs Biowulfe broga gecyðed
snude to soðe, þæt his sylfes ham,
bolda selest, brynewylmum mealt,
gifstol Geata. ...
Hæfde ligdraca leoda fæsten,
ealond utan, eorðweard ðone
gledum forgrunden....

Then the spirit began to spit fire, to burn the bright homes... He surrounded the land-dwellers with fire, flames and torches.... Then the truth of that horror was told quickly to Beowulf, that his own home, best of buildings, gift-house of the Geats, was consumed in the waves of fire.... The fire-breathing dragon had destroyed the homes of the people, islands to the outside and the fortified town, with burning coals. (ll. 2312-13a, 2321-22a, 2321-27a, 2333-35a).

The foreshadowing of Heorot's destruction is brought to closure by the burning of Beowulf's castle. Unlike for Heorot, however, there is no specific description of the homes, castles, and fortresses razed by the dragon in his fiery fury. The people are 'surrounded by fire,' their homes apparently destroyed around them. In addition, while it is foreshadowed that Heorot will destroyed by human actions – 'ecghete' ('war,' l. 84) and 'wælniðe' ('hostility,' l. 85), the destruction wrought by the dragon is something else, a force neither human nor natural.

From an ecocritical point of view, it might be argued that the violence of humans is displaced onto monstrosity, so that human responsibility can be deflected. But the dragon's destruction also resembles the fury of the storm described in *Exeter Book*, Riddle 1, in which the storm declares, 'ic... folcsalo bærne, / ræced reafige' ('I... burn a public hall, lay waste to a palace,' ll. 5-6). The dragon's rage might be understood as a cultural effort to make sense of natural forces that sometimes result in violence against humans as well as destruction to the natural world itself. The divine power depicted as affecting the non-human natural world in *Genesis A* and *Exodus* here becomes something more demonic. The ascription of natural forces of destruction to divinity or monstrosity, in either case beyond the categories of human or natural, has the effect of rejecting the possibility that nature possesses an agency independent of human interests.

In the narrative of the destruction wrought by the dragon in *Beowulf*, there is no reference to the destruction of agricultural land or domestic animals, in contrast to the *Exeter Book*, Riddles 1, 2, and 3, about storms. The riddles mention the destruction of homes and palaces, but also describe trees being felled, waves whipped up, and earth disturbed. The focus in *Beowulf* on the destruction of built dwellings, specifically associated with human occupants, is consistent with the focus in the poem on human needs and desires; the contrast with the *Exeter Book* riddles is also consistent in that the latter give more attention and grant greater moral agency to plants and to animals, wild as well as domesticated. (See Chapter 6 for an extended discussion of the 'Storm' riddles.)

The stormy destruction that is described in the *Exeter Book* riddles contrasts with human and dragon destruction in *Beowulf*. The poem foreshadows the burning of Heorot as a result of human hostilities in the first third of the poem; the dragon's fiery destruction in the poem's final third is called forth by human actions, after the dragon has long rested peacefully on its hoard. Yet when Wiglaf finally sees the inside of the dragon's lair, he sees treasure that has suffered from time and the absence of human attention: the drinking-vessels of departed men lack polishers ('feormendlease,' l. 2761); and 'þær wæs helm monig / eald ond omig' ('there was many a helmet, old and rusty,' ll. 2762b-63a). These details recall *The Ruin*, fallen into decay because it lacks people to repair it. The dragon, which wreaks ruin on Beowulf's home and those of his people, lives in a barrow that has been ruined by time and the forces of nature.

Grendel's threats to Heorot, and the dragon's destruction of Geatland, are not human, yet they are also not natural: as noted above, deer in the forest will allow themselves to be shot by hunters rather than approach Grendel's lair. Yet the ultimate threat to Heorot comes from human hands, not from a destructive or hostile natural world. Twenty-first century environmentalists see new threats to global stability in the human actions leading to climate change; the newly felt sense of threat and impermanence is often assumed to be in contrast to the relative environmental integrity of earlier periods. Ecocritics feel a new urgency about impermanence around places in which we live, arguing for a newly archipelagic, disconnected understanding of our dwelling places in a time of increasing travel, migration across and among continents, and the construction of mass-market 'non-spaces' (Buell 2005: 69-70) such as fast-food joints and airports, indistinguishable one from another. But already in *Beowulf* there is a sense of place as impermanent, threatened not by natural forces but by human acts of war.

Beowulf closes with the construction of a final edifice: a grave for the remains of Beowulf, and a monument to his memory:

> Geworhton ða Wedra leode
> hleo on hoe, se wæs heah ond brad,
> wægliðendum wide gesyne,
> ond betimbredon on tyn dagum
> beadurofes becn....
>
> Then the Wether-Geatish people built a shelter on the heights; it was high and broad, seen widely by sea-farers, and they built for ten days a beacon for the war-bold one. (ll. 3156-60a)

The monument is meant to last for a long time, to be both useful and sentimental, to celebrate and remember the dead man and, as Beowulf hopes, to serve as a navigational beacon for future seafarers. His barrow contrasts with the sea-burial of Scyld, with which the poem opens, and of the dragon: both Scyld and the dragon disappear beneath or beyond the waves, while Beowulf wishes for his barrow to be seen and his name remembered. Hell and Schönle argue that ruins 'reveal an ambivalent sense of time, at once the awareness of an insuperable break from the past... and the sense that some valuable trace has endured' (5). *Beowulf* describes such a break between the pagan past and the Christian present combined with the impulse to preserve elements of that past. The fragments of the past preserved in *Beowulf* are varied and complex, including details about the historical Hygelac and the legendary *Beowulf* as well as about the landscape of the country from which the people have migrated and the alliterative epic verse itself, written down in preservation of an oral tradition also on the verge of threat from Norman hostilities which would lead, a generation or two later, to great cultural change.

Alfred Hiatt argues that in *Beowulf*, 'identity derives from interpersonal relations rather than from a relationship to land. Consistent with other early medieval expressions, then, the idea of nation resides with a people, and not with a spatially defined entity' (32). A notion of dwellings as temporary, alongside the evidence of abandoned Neolithic and Roman constructions across Anglo-Saxon England, is consistent with a non-geographic definition of identity during the period. The focus on ruins rather than on usable dwellings expresses the sense of geographical belonging as temporal and temporary, and also emphasizes the Anglo-Saxon sense of identity as unstable and changeable. Simultaneously, it expresses a Christian insistence on earthly existence as a brief span within the more important idea of eternal life. As Beasley-Murray argues 'Ruins demonstrate that whole cultures, just like the lives of mortals, are transient. Hence they are invented by cultures that feel their own transience' (213). The fragility of human edifices and political entities in *Beowulf* allows for a multiplicity of readings.

Conclusion

Julia Hell traces to the writing of Scipio in 146 BC a shift from 'the *vanitas* or *memento mori* trope to a discourse concerned with the inevitable rise and fall of empires' (170). The poems discussed in this chapter all deal with the rise and fall of civilizations in the past, with the recognition implied that their own social order was also threatened – and in fact during the ninth and tenth centuries the Danes regularly attacked England. The *Exeter Book*, the Junius manuscript, and the Nowell codex, containing *The Ruin*, *Genesis A* and *Exodus*, and *Beowulf*, respectively, were all written down near the end of the tenth century; the poems may have been composed earlier, but were likely shaped and certainly chosen for preservation during a time when the *Anglo-Saxon Chronicle* entries repeatedly recorded the burning of monasteries and towns.

The Ruin is destroyed because of the absence of repairers. In *Genesis A*, the people of Sennar attempt to construct a tower as a sign of their power within a city, but they are frustrated by divine intervention; the Tower of Babel remains half-finished because those who would have built it are scattered. *Exodus* imagines a foundation of ruins within the sea for a society that will build anew in Israel as well as in Anglo-Saxon England: the Israelites are, in Christological interpretation, seen as the pre-figurations of Christianity; in Bede's interpretation, the Anglo-Saxons have taken on Christian worship and come to view themselves as the new Israel. Beowulf wants his barrow to be both *memento mori* and a thing of utility, a signpost to sailors to aid in navigation. The poem ends with Beowulf's death and the anticipated collapse of his kingdom, and this conclusion signals the end of pagan Danish civilization with both geographic and religious transitions into Christian Anglo-Saxon England.

The descriptions of ruin in these poems preserve them in memory. Landscapes are inscribed and over-inscribed with literal and historical and religious meanings that convey a variety of orientations to nature and of human relationships to the natural world. Landscapes are sometimes imagined as static backgrounds to human concerns; in other parts of the same poems, they become mutable and unsettled environments in which humans move at their own risk. The examples of *The Ruin*, *Genesis A*, *Exodus*, and *Beowulf* suggest that the Anglo-Saxons feel their relationship with their environments, whether natural or built, is transient. They remember, if dimly, a migration from another place and a transition from pagan to Christian. When the Anglo-Saxons write about or imagine ruins, distant in time and/or in place, they also comment on their own cultural preoccupations about the transitory nature of individual life as well as of tribes, kingdoms, and cultures.

In response to Ursula Heise's acknowledgement that attachment to place is neither necessary nor sufficient for contemporary ecological engagement, Axel Goodbody argues 'Literary remembering ... can also be understood as a *performance* of the past.... Historical facts can be readily invested with new meanings' and that actual remembered places have 'overlapping, often conflicting significances, and [are] open to constant reinterpretation and reappropriation' (58, 60). Goodbody argues that in fact environmentalists need to attend to place, but should recognize that attachment may be to remembered, imagined, culturally constructed notions of place, rather than transparently perceived or remembered ones. The attention to ruins in places temporally and/or geographically distant in the Old English poems considered here suggests that Anglo-Saxon attachments to and understanding of place shifted and interacted in complex ways.

In the next chapter, I turn to the Latin and Old English lives of Guthlac. The Old English saint withdrew to a hermitage in the East Anglian fenlands where he spent the last several years of his life in the early eighth century. The island where Guthlac made his home contained ruins of former use and possible habitation. While the Anglo-Saxons adopted Hebrew and Christian traditions and imagined themselves in continuity with that past, Guthlac's biographers instead erased from existence the prior inhabitants of his refuge, re-imagining the place as a wilderness and its former occupants as demons. Following his career as a soldier in the Mercian wars against Welsh and Britons, Guthlac's literal occupation of these ruins, and their figurative use in the narratives of his sainthood, enact rupture rather than continuity. Felix's *Vita* and its Old English translations and adaptations articulate a justification for appropriation of other peoples' lands, suggesting that Guthlac's Christian faith gives him dominion over the earth and animals and other people, anticipating the colonizing actions and rhetorics of much later English culture.

4 Rewriting Guthlac's Wilderness

Introduction

In turning from ruins to wilderness and the Old English narratives of Guthlac, I continue an exploration of how Anglo-Saxon literary and documentary texts constructed lived places and the spaces contiguous to them, and what cultural contexts those constructions reveal. Saint Guthlac withdraws to an island containing a ruined structure, once a grave, later a cistern; I discuss that in this chapter rather than with the previous examples of ruins because of the way the ruin intersects with constructions of wilderness and the people who live in supposed wildernesses, as well as with the ways in which the Guthlac narratives anticipate much later colonizing invasions.

The early English saint Guthlac lived from 674 to 714, the last fifteen years as a hermit on an island in the East Anglian fenlands. At the request of King Ælfwald of the East Angles, the monk Felix wrote about Guthlac's life in Latin between about 730 and 740, relying on accounts of Guthlac's successor in the hermitage, Cissa, as well as of his frequent visitor Wilfred, among others (*Felix's Life* 6, 15). Felix frequently borrows details from Bede's *Life of St. Cuthbert* and Gregory's *Life of St. Boniface*, as well as from other saints' lives (*Felix's Life* 16-17). Felix's *Life* was translated into Old English prose, with versions of the material in two different manuscripts of the eleventh century. It was also adapted into two different poems, both preserved in the *Exeter Book*, a significant collection of Old English poetry written near 1000; in the later Middle Ages, it was translated and adapted again. An abbey, of which only ruins survive today, was built on the site of Guthlac's hermitage.

In thinking about the versions of Guthlac's life, this chapter draws on the insights of postcolonial theory as well as of ecocriticism, though there are potential pitfalls in thinking through the life of a member of a majority culture though a postcolonial lens. In a discussion of colonial expansion in Canada and Australia that draws on ecocritical insights, Kylie Crane writes:

> The use of the term postcolonialism in conjunction with a project that focuses only on settler colonies and, further, only on the literature produced by authors from perspectives that are best described as settlers – that is, from the privileged – may be troubling (19).

The same might be said of an analysis of Felix's *Vita* and the Old English prose and poetic adaptations: works by and about people privileged in

Anglo-Saxon England: a member of the royal family of the Germanic people that conquered the Britons, a high-ranking cleric who becomes a monk, as written about by another monk, a Christian male of the dominant religion. Yet thinking about this text through postcolonial theory makes it possible to uncover some troubling things: namely, the ideas that enable later colonial expansion are already present in Anglo-Saxon texts, in both the Anglo-Latin original and the Old English translation and adaptations.

Post-colonial criticism attends to power relationships and knowledge imbalances between colonizers and colonized. Post-colonial theory usually focuses on modern empires, on colonies or former colonies, and their subjects, on the notion of the 'other,' on who speaks on whose behalf. But it can also help to understand what happened when Angles and Saxons invaded Britain in the years following the Roman withdrawal and supplanted the native Britons in positions of influence and power. The *Lives* of Guthlac present that overthrow, as well as the subsequent battles between the 'englisc,' and the Welsh and Britons who continued to occupy the margins of Britain in Wales in the west as well as in the fens in the east, as legitimate, and in fact as being in no need of any justification. Attending to the ways in which a text like the *Life of Guthlac* normalizes and naturalizes conquest and power imbalances illuminates the development of a world-view that enables later colonizing operations to be undertaken as if it were legitimate for the English to subdue the Irish and, later, the inhabitants of North America, China, India, and Africa.

Postcolonial Ecocriticism

Postcolonial ecocriticism attends to the intersections between colonization and environmental degradation, to issues of environmental justice, and to the ways in which people of color are aligned with the earth in a paradigm that makes a parallel between 'natural resources' and 'human resources,' thus placing some humans, alongside animals and the environment, in a position of material utility to other humans, usually white and male. Laura Wright cautions that in looking at intersections between postcolonial and environmental theories, 'it is necessary to examine the ways that environmentalism, as the social movement that gave birth to the kinds of ecocritical analysis codified by Buell and other scholars, is a Western concept' (10) and that 'post-colonial ecocriticism – particularly in its current manifestation as a white and, more often than not, Western academic discourse – is further complicated by its need to maintain rigorous and sustained literary critique while simultaneously working to avoid speaking for the environmental needs of non-Western people and landscapes' (175). As a white, tenured faculty member at a North American university, albeit neither male nor Christian, I write from a position of substantial privilege, and undoubtedly with blind spots. In thinking and writing about occupants of England and Wales of the distant past, and their subjugation by invaders, from perspectives developed by modern scholars of colonialism, I attempt to tread carefully as I seek to understand how the Anglo-Saxons' views toward people outside of their own communities intersected with their view of the land and animals, positioning both as available literally, ideologically, and spiritually for the use of those in the majority culture.

This chapter does not explicitly address ecofeminism, though it is informed by the ideas discussed in Chapter 2 with respect to Grendel's mother and Chapter 6 with respect to the *Exeter Book*, Riddle 52 which depicts a female slave engaged in manual labor, or supervising criminals, and simultaneously as a metaphorical stand-in for an object. Women are absent from the *Vita Guthlaci* and its Old English translation and adaptations, except as objects of rape, much as they are absent from *Andreas*, as discussed in Chapter 2, as well as from the Old English poetic *Exodus*. But the ecofeminist insight that European culture systematically privileged men over women helps to understand how it also privileged some men over all other humans.

Felix's *Vita Guthlaci* and the Old English versions of the *Life of Guthlac* locate Guthlac's hermitage in the 'wilderness,' then as now a troubled concept. In this chapter, I discuss Felix's Latin *Vita* alongside the anonymous Old English prose translation as well as the poetic *Guthlac A*. As Kylie Crane writes, 'wilderness entails a colonial gesture, placing indigenous presences and practices under erasure' (Crane 2). Women are likewise erased from the narrative of Guthlac's life. Crane points out that wilderness is 'marked by its natural qualities' and 'conceived in terms that oppose it to civilization' (14-15). This description of wilderness serves aptly for the spaces that Felix and the anonymous Old English prose translator and adapting versifiers deploy. In the 'wilderness' he occupies, however, Guthlac encounters demonic Britons or British demons. This anticipates later colonizing incursions in which the land occupied by native peoples is understood as a wilderness, so as to enable discounting their occupation of and right to it – or, less passively, that this discourse participates in a view of English expansion that enabled later colonizing violence.

The importance of activism links postcolonial and environmental critiques, as Graham Huggan and Helen Tiffin argue in their book *Postcolonial*

Ecocriticism, and both are threatened by multinational corporations. Post-colonial, ecocritical scholarship 'finds itself increasingly compromised by a global capitalism that has not always been challenged to the same degree as the imperial behaviors it instantiates and inspires' (11). An effective challenge to eco-imperialism requires a rejection of consumer culture. The economic system of Anglo-Saxon England did not encourage excessive consumption on the scale of contemporary developed societies, and in fact one of the things that characterizes many hermits, including Guthlac, is the ascetic renunciation of material goods and comforts. But the assumptions enabling Guthlac's move into a hermitage, as well as his biographer's assertions that plants and animals existed to serve him, also underscore modern capitalist ideas that animals and landscapes are limitless resources available for human use. As Huggan and Tiffin further argue 'what the postcolonial/ecocritical alliance brings out, above all, is the need for a broadly materialist understanding of the changing relationship between people, animals, and environment – one that requires attention, in turn, to the cultural politics of representation' (12). Understanding how the ideologies that underpin capitalism and colonization already suffused the life of an eighth-century saint at the margins of Christianity can help to illuminate how those same ideologies continue to operate today and how they have evolved to the point of being a threat to planetary stability.

Post-colonial ecocriticism foregrounds environmental justice, noting how imperial conquest often depends upon disruptive and destructive mining or drilling into the earth, replacement of subsistence farming with cash crops, and the abuse of human labor in slavery or indentured servitude. Huggan and Tiffin point out that it can be hard to define either postcolonial or ecocritical theory, but they locate a point of convergence in the work of the ecofeminist philosopher Val Plumwood. Plumwood pointed out that Greek and Roman culture, as well as Jewish and later Christian traditions, defined humans in association with 'reason' and in opposition to 'nature,' excluding from moral consideration beings without reason – a category that included animals and the earth and also ended up encompassing women as well as 'othered' people. Plumwood calls this paradigm 'hegemonic centrism' (2002: 148), and asserts that it has enabled racism, sexism, and colonialism, along with environmental degradation. Huggan and Tiffin note: 'as Plumwood argues, the western definition of humanity depended – and still depends – on the presence of the "not-human": the uncivilised, the animal and animalistic. European justification for invasion and colonisation proceeded from this basis, understanding non-European lands and the people and animals that inhabited them as "spaces", "unused, underused or empty"' (5). The feminine is explicitly 'other' in the *Vita Guthlaci*, implied only in reference to rape as object of Anglo-Saxon violence, and otherwise completely absent. The Britons who fought against Anglo-Saxon occupation are depicted as non-human in their association with the demons who harass Guthlac after he retreats to his hermitage.

Huggan and Tiffin follow Anthony Vital on the importance of language as a point of intersection between postcolonial and ecological criticisms: 'Vital suggests that the best way to reconcile postcolonial criticism and eco/environmental criticism might be to take into account "the complex interplay of social history with the natural world, and how language both shapes and reveals such interactions" (90)' (Huggan-Tiffin 15). In carefully reading the Latin and Old English *Lives* of Guthlac, I seek to uncover some of the ways in which language reveals interactions between the natural world and previous occupants of a place conceptualized as 'wild,' on the one hand, and a person privileged by class, religion, and gender, on the other. By constructing the landscape and its occupants in these ways, the Guthlac narratives – which were adapted and translated in Middle English versions as well, thus reaching nearly the point of actual colonial expansion – became part of a textual tradition that shaped interactions with colonized peoples and lands. For the Anglo-Saxons this was a distant future, but it was a future that Anglo-Saxon cultural artifacts helped to enable.

The site of Guthlac's hermitage is characterized as a 'wilderness,' an early iteration of a long tradition of defining the land of 'other' people as uninhabited, essentially by making the claim that such other people are not fully human. As Huggan and Tiffin point out, 'throughout western intellectual history, civilisation has consistently been constructed by or against the wild, savage, and animalistic, and has consequently been haunted or "dogged" by it' (Huggan-Tiffin 134). The *Lives* of Guthlac are assumed to 'describe' the wilderness of the fens, but I argue that in fact they imagine and construct it, first insisting that the fens *are* untracked and uncultivated despite evidence to the contrary, then imagining the creatures that occupy the area around Guthlac's hermitage as demonic rather than human. Moreover, the Old English poetic version of Felix's Latin *Vita*, *Guthlac A*, re-imagines and re-constructs that wilderness in ways that depart substantially from the Latin or the prose translation.

Crane argues that 'wilderness' is continually used and re-imagined in opposition to 'civilization', and as 'civilization' expands, the meanings and territories designated as 'wilderness' are in on-going flux. She argues that reading through both ecocriticism and postcolonial theory 'enable[s] the development of an informed reading position that queries the assumptions of wilderness engendered in the respective narratives.... Whether considered semantically or politically, wilderness clearly remains a highly contested space' (12). What is meant by 'wilderness' has to be considered within political, linguistic, and cultural contexts and subtexts. In the *Lives* of Guthlac, 'wilderness' is presented as uncontested space, even as the language that describes it evokes the ghostly traces of previous habitation. The island on which Guthlac builds his hermitage is 'untracked' and 'uncultivated' and known only to Tatwine – but then, it *is* known to Tatwine, and it turns out to be tracked after all. Guthlac fights off repeated attacks from 'demons' as he establishes his residence in his hermitage. Constructing the space as 'untracked' allows Felix as well as the translator and the versifiers to imagine the occupants who attack Guthlac as 'demons,' attempting to erase prior human presence by redefining it as not human.

Guthlac as Warrior

Felix's prose *Vita Guthlaci* opens with an account of Guthlac's ancestry through his father Penwalh down to Icel. In the *Anglo-Saxon Chronicle* entry for 755, Icel's ancestry is traced further back through Offa to Woden (*Two Saxon Chronicles* 50). Felix locates Guthlac's ancestry in an important line in literary, historical and mythical terms. (The later Offa, King of Mercia, had not yet come to power at the time when Felix wrote.)

Felix describes Guthlac as an obedient, affectionate, even-tempered child, but as he reaches adulthood, his temperament shifts:

> iuvenili in pectore egregius dominandi amor fervesceret, tunc valida pristinorum heroum facta reminiscens, veluti ex sopore evigilatus, mutata mente, adgregatis satellitum turmis, sese in arma convertit.
> A noble desire for command burned in his young breast, he remembered the valiant deeds of heroes of old, and as though awakening from sleep, he changed his disposition and gathering bands of followers took up arms. (80, 81)

The Old English translation makes some intriguing changes to the Latin:

> þa gemunde he þa strangan dæda þara unmanna and þæra woruldfrumena; he þa, swa he of slæpe onwoce, wearð his mod oncyrred, and he gesomnode miccle scole and wered his geþoftena and hys efen-hæfdlingas, and him sylf to wæpnum feng.

> Then he remembered the mighty deeds of the *unmanna* and those of days of old; then, as if he awoke from sleep, and his mind changed direction, and he gathered many fellow pupils and armed his companions and his equals, and took weapons to himself. (*Anglo-Saxon Version* 12, 14)

The meaning of 'unmanna,' untranslated in the passage above, is unclear, possibly pointing to 'heroes,' possibly 'evil ones.' The Old English version, unlike the Latin, suggests that Guthlac gathered his schoolmates to go to battle, emphasizing this by reference both to 'scola' and to 'efen-hæfdlingas,' a repetition with no parallel in the Latin. Moreover, the Old English translator renders the Latin 'heroum,' meaning 'hero' or 'demigod,' with another double construction, using both 'unmanna' and 'woruld-frumena.' The word 'unmann,' 'evil one, demon' might be an error for 'iumann,' 'man of old,' as suggested by the use also of 'woruld-frumena.' But 'woruld-frumena' is itself an odd compound, perhaps emphasizing that the companions Guthlac gathered were men of the world rather than the angels who would later become his companions.

Alfred Siewers argues that *Guthlac A* was written, like Felix's *Vita*, in the eighth century, at a time when 'the cultural identity of Anglo-Saxon Mercia as a kingdom and as a people was still a work in progress in the eighth century … and this has important implications for the literary landscapes associated with it and their related religious ideology' (2003: 8). *Guthlac A* is preserved only in the late tenth-century manuscript known as the *Exeter Book*, and there is little internal evidence to determine its initial date of composition. Felix's *Vita*, however, can more reliably be dated by external evidence to the early eighth century, so while Siewers' analysis may not be applicable to the poem, it surely applies to the Latin *Vita*. For the Latin text, Guthlac's battles with Welsh warriors in his early career as warrior, echoed by the later battles with demons, represent a cultural and military clash from living memory. Contemporary historical accounts tell us that the Mercians shared a border with Wales, and that the Welsh and Mercians engaged in frequent battles. The building of a massive earthwork known as Offa's Dyke is attributed to the Mercian King Offa, a half-century later, but recent archaeological evidence suggests the possibility that construction began earlier ('Could Offa's Dyke' 10).

The account of Guthlac's battles with enemy troops begins with the narrative of his participation in plundering the bodies of the enemy dead, before moving to a description of the actual fighting. Felix writes that Guthlac always gives back part of the spoils of war, in a kind of apologetic acknowledgement that burning, killing, and taking contraband are perhaps not ideal activities for a future saint:

> Et cum adversantium sibi urbes et villas, vicos et castella igne ferroque vastaret, conrasis undique diversarum gentium sociis, inmensas praedas gregasset, tunc velut ex divino consilio edoctus tertiam partem adgregatae gazae possidentibus remittebat.
> But when he had devastated the towns and residences of his foes, their villages and fortresses with fire and sword, and, gathering together companions from various races and from all directions, had amassed immense booty, then as if instructed by divine counsel, he would return to the owners a third part of the treasure collected. (80, 81)

Plundering booty from one's dead and defeated enemies was a necessary feature of the early Anglo-Saxon economy, but in literary texts it is frequently glossed over or attributed rather to the enemy than the heroes. The emphasis on Guthlac's return of a third of the collected loot echoes this sense that there is something less than noble about the activity.

The Old English version of the *Vita* is more direct about the negative connotation of looting. The account of taking the loot is briefer, saying simply that Guthlac 'of mannum heora æhta nam' ('took their possessions from men,' 14). This is immediately followed by divine reproach, in a passage with no parallel in the Latin:

> Þa wæs he semninga innan manod godcundlic and læred þæt he þa word hete, ealla þa he swa genam he het þriddan dæl agifan þam mannum þe he hit ær ongenæmde.
> Then he was suddenly inwardly divinely admonished and instructed that he must then give the command, all that which had thus taken, he commanded the third part to be given back to the men from which he had earlier taken it. (14)

Moreover, there is a shift in responsibility. While the Latin Guthlac returns the plunder himself, the Old English Guthlac commands his men to return it. Transferring the task of returning the loot to his men perhaps dilutes Guthlac's responsibility for having done the deed in the first place.

Having made excuses for Guthlac's pillaging of the enemy, Felix goes on to recount his victories during his nine years as a warrior, during which

> persecutorum suorem adversantiumque sibi hostium famosum excidium crebris vastationum fragoribus peregisset, tandem defessis viribus post tot praedas, caedes rapinasque quas arma triverunti, lassi quieverunt....

> He had achieved the glorious overthrow of his persecutors, foes and adversaries by frequent blows and devastations, at last their strength was exhausted after all the pillage, slaughter, and rapine which their arms had wrought, and being worn out, they kept the peace. (80, 81)

Guthlac's 'glorious overthrow' of his enemies, and his return of a third of the spoils of war, apparently excuses rape, killing, and looting, as well as the burning of houses and presumably fields. Domesticated animals were frequently taken from the defeated foes, along with stored food, weapons, armor and shields, and other valuables, perhaps including coins. Peace finally comes because the enemy is 'exhausted' by pillage, assault, death and destruction. The lack of any kind of comment on the idea that sexual assault of women can be a tool of warfare makes rape appear 'normal' alongside burning and killing and pillaging, and the implication is that restitution is as simple as returning a portion of the plunder. When burning villages, Guthlac likely destroyed fields as well as houses: domesticated as well as wild animals and plants are killed alongside the human occupants. The land occupied by the human enemy itself becomes the target of Guthlac's violence. Given the references in the *Exeter Book* riddles to female Welsh slaves (see Chapter 6), Guthlac's loot may have included humans taken with the express purpose of enslaving them.

The Latin continues with a sustained meteorological metaphor, saying that Guthlac 'beatae memoria' ('of blessed memory,' 80, 81) spent these years tossed by storm and waves among gloomy clouds. The passage uses details of the storm in an instrumental sense, subordinating them to human concerns; see Chapter 6 for a discussion of the 'Storm' riddles (1-3) in which weather events are described on their own terms rather than as symbolic of human concerns. The Old English translation radically abbreviates the passage, recording that 'se eadige Guthlac' ('the blessed Guthlac') spent nine years in 'ehtnysse' ('persecution') of the enemy, during which 'he hine sylfne betweox þises andweardan middaneardes wealcan dwelode' ('he himself dwelt among the presence of the waves of middle-earth,' 14). The metaphor of storm and the discussion of Guthlac's warrior activities are reduced, and the reference to the 'divine overthrow' of the enemy is completely eliminated. The effect is to lay greater emphasis on the divine admonition to return the property he has pillaged from those he had defeated.

The text moves from the descriptions of battles to a description of a night in which Guthlac contemplates 'antiquorum regum stirpis suae per transacta retro saecula miserabiles exitus flagitioso vitae termino' ('the wretched deaths and shameful ends of the ancient kings of his race in the course of the past ages,' 82, 83). The Old English translation follows the Latin quite closely here: 'he geþohte þa ealdan kyningas þe iu wæron, þurh earmlicne deað and þurh sarlicne utgang þæs manfullan lifes, þe þas woruld forleton' ('he thought about the old kings that used to be, who abandoned this world through miserable deaths and through lamentable departures from their evil lives,' 14). Both texts use the battle against the enemy to illustrate Guthlac's development through young adulthood to his decision to dedicate himself to serving God by becoming a hermit. The people who have been defeated, the homes and fields and crops and domesticated animals destroyed, and the individuals killed are not important in the narrative of Guthlac's life. His Christian faith is what matters.

Women as objects, actual objects, animals, cultivated fields and dwellings, people of other social groups: all become analogous objects in warfare. Postcolonial ecocriticism attends to the intersections between the treatment of humans and the non-human; postcolonial feminism points to the intersections between gender and race. Attending to intersections among them reveals how the Old English versions of *Guthlac* demonstrate that the ideologies that enabled the subjugation of humans and animals and landscape to the needs and desires of European colonizers in India, Africa, China, and the Americas were not Renaissance inventions, but extensions of ideas articulated and developed in Anglo-Saxon England.

Guthlac as Hermit

Huggan and Tiffin point out that while European settlers occasionally adopted ideas from the cultures they inhabited, they were more likely to import their own ideologies. 'More usually ... ideas of animal treatment and land use initially formed in Europe predisposed colonial administrators and settlers to a facile belief in the apparently limitless resources of the settler colonies' (8). Felix's *Life of Guthlac* and its old English translation and poetic adaptation, itself part of a textual tradition of saints' lives, transmits ideas from Latin literate culture as well as Anglo-Saxon warrior society; they show the Anglo-Saxons ignoring all inhabitants of a territory they wished to colonize (whether animals or humans) and calling it 'weste,' a word with a range of definitions including 'waste' as well as 'desert' or 'wilderness.' Huggan and Tiffin continue 'Such places, after all, were apparently untamed, unowned and, above all, unused; and, accordingly, settlers set about rendering them productive and profitable through imported methods rather than by accommodating them to local circumstances' (8). Guthlac and his biographers deny the evidence of prior human habitation in their characterization of the fenland island where he makes his hermitage as wilderness.

Having decided to become a hermit, Guthlac sets off to find a place within this fen to live. The passage rewards quotation and examination at length:

> Est in meditullaneis Brittanniae partibus inmensae magnitudinis aterrima palus, quae, a Grontae fluminis ripis incipiens, haud procul a castello quem dicunt nomine Gronte, nunc stagnus, nunc flactris, interdum nigris fusi vaporis laticibus, necnon et crebris insularum nemorumque intervenientibus flexuosis rivigarum anfractibus, ab austro in aquilonem mare tenus longissimo tractu protenditur. Igitur cum supradictus vir beatae memoriae Guthlac illius vastissimi heremi inculta loca conperisset, caelestibus auxiliis adiutus, rectissimo callis tramite tenus usque perrexit.
>
> There is in the midland district of Britain a most dismal fen of immense size, which begins at the banks of the river Granta not far from the camp which is called Cambridge, and stretches from the south as far north as the sea. It is a very long tract, now consisting of marshes, now of bogs, sometimes of black waters overhung by fog, sometimes studded with wooded islands and traversed by the windings of tortuous streams. So when this same man of blessed memory, Guthlac, had learned about the wild places of this vast desert, he made his way thither with divine assistance by the most direct route. (86, 87)

The Old English version follows the Latin fairly closely, though there are several differences:

> Ys on Bretone-lande sum fenn unmaetre mycclnysse þæt onginneð fram Grante ea naht feor fram þære cestre, ðy ylcan nama ys nemned Granteceaster. Þær synd unmaete moras, hwilon sweart waeter-steal, and hwilon fule ea-riþas yrnende, and swylce eac manige ealand and hreod and beorhgas and treow-gewrido, and hit mid menigfealdan bignyssum widgille and lang þurhwunað on norð-sæ. Mid þan se foresprecena wer and þære eadigan gemynde Guðlac ðæs widgillan westenes þa ungearawan stowe þær gemette, þa wæs he mid godcunde fultume gefylst, and þa sona þan rihtestan wege þyder togeferde.
>
> There is in Britain a fen of immense size, which begins from the river Granta not far from the city, which is named Grantchester. There are

> immense marshes, now a black pool of water, now foul running streams, and also many islands, and reeds, and hillocks, and thickets, and with manifold windings wide and long it continues up to the north sea. When the aforesaid man, Guthlac of blessed memory, found out this uncultivated spot of the wide wilderness, he was comforted with divine support, and journeyed forthwith by the straightest way thither. (20, 21)

The Old English version eliminates the characterization of the fen as 'most dismal' while emphasizing the 'immense' size of the region with a doubled use of the word 'unmæte' to modify both 'fenn' and 'moras.' The detail from the Latin *Vita* that the islands are wooded 'nemorum' is shifted in the Old English, which instead makes reference to 'hreod' and 'treow-gewrido,' 'reeds' and 'thickets.'

A man named Tatwine takes Guthlac by boat 'per invia lustra' ('through trackless bogs') to 'insula media in palude posita quae ante paucis propter remotioris heremi solitudinem inculta vix nota habebatur' ('an island in the middle of the marsh which on account of the wildness of this very remoted desert had hitherto remained untilled and known to a very few,' 88, 89). Colgrave's translation of 'invia lustra' as 'trackless bogs' is only one possibility for the phrase; it could also be rendered as 'impassable wilderness.' For an ecotheoretical analysis, the differences are not inconsequential. 'Impassable' suggests that it is not possible to move through the area, though the fact that Tatwine is taking Guthlac through it contradicts this. 'Trackless' thus seems the more logical translation, though this too is contradicted by the evidence revealed in the following paragraphs about prior habitation. In the Old English, Tatwine takes Guthlac on his boat 'þurh þa rugan fennas' ('through the uncultivated fens'); 'wæs þæt land on middan þam westene' ('the land was in the middle of the wilderness').

Both Felix and the Old English translator imagine the location of Guthlac's hermitage specifically as having no prior human activity, though the language they use to convey this varies slightly. In both versions, the fact that the land is 'inculta' or 'rug,' i.e. 'uncultivated' is taken as synonymous with its status as 'lustra' or 'westen,' i.e. 'wilderness.' Both versions, in other words, assume that agricultural cultivation is synonymous with occupation of a landscape. We know today that hunting-and-gathering societies have existed for many millennia, and it appears that those who lived in the East Anglian fens survived in just such a fashion, because the frequently-flooded marshes could not be used to plant crops. Leo Mellor argues that 'to imagine a landscape as a wilderness has often been, throughout history, a way to render it into a *tabula rasa* for the imagination, thronged with natural forces and ripe with possibilities – but scythed clear of human presence. This is mendacious when not merely naïve' (Mellor 111). It is tempting to excuse the Anglo-Saxons as being simply unaware that fen-dwellers existed, but they must have known from travel and trade that they did. Another possible explanation for the description of the area as 'wasteland' or 'wilderness' is that Guthlac, Felix and their contemporaries could not understand the fenland people as human. Such a mindset enables Guthlac's occupation of the island and sets a precedent for much later colonizing violence.

Felix tells his readers that after spending some days on the island, Guthlac determines to return to the monastery and bid farewell to his companions there; after ninety days, 'ad supradictum locum, quasi ad paternae hereditatis habitaculum ... regressus est' ('he returned to the above-mentioned place whence he had come, as though to a home inherited from his father,' 90, 91). In the Old English, this is rendered as 'to þære stowe þæs leofan westenes' ('to the place of his beloved wilderness,' 22). It is interesting that the Old English text (but not the Latin) calls Guthlac's hermitage 'beloved' on account of its location in the wilderness.

At this point, Felix provides additional details about the island. Despite the earlier insistence that the land had been untracked and unused, it turns out that it had seen prior human habitation:

> Erat itaque in praedicta insula tumulus agrestibus glaebis coacervatus, quem olim avari solitudinis frequentatores lucri ero illic adquiriendi defodientes scindebant, in cuius latere velut cisterna inesse videbatur; in qua vir beatae memoriae Guthlac desuper inposigo tugurio habitare coepit.
> Now there was in the said island a mound built of clods of earth which greedy comers to the waste had dug open, in the hope of finding treasure there; in the side of this there seemed to be some sort of cistern, and in this Guthlac the man of blessed memory began to dwell, after building a hut in it. (92-94, 93-95)

The Old English text is quite similar to the Latin at this point:

> Wæs þær on þam ealende sum hlaw mycel ofer eorðan geworht, þone ylcan men iu geara for feos wilnunga gedulfon and bræcon. Þa wæs þær on oþre sidan þæs hlawes gedolfen swylce mycel wæter-seað wære. On þam seaðe ufan se eadiga wer Guthlac him hus getimbrode, sona fram fruman þæs þe he þæt ancer-setl gesæt.
> There on the island a great mound had been constructed on the earth; that same (mound) men in former years had dug and broken up in desire of treasure. There on the other side of the mound a hole had been dug

> up, that was like a large well. Above the well, the blessed man Guthlac built himself a house, immediately from the beginning that he dwelt in that hermitage. (26)

Both texts refer quite clearly to several layers of pre-existing occupancy: the people who initially buried the treasure; the people who dug in search of that treasure later on; and the people who constructed a cistern in the side of the mound. O'Brien O'Keeffe summarizes the evidence for occupation of the Anglian fenland: 'There is evidence of roman colonization on the silts, its retraction in the later second century after flooding, and possible rebuilding in the fourth century, but the archaeological record of settlement is difficult and complicated by silting. The evidence of settlement in the early Anglo-Saxon period is thus ambiguous in the extreme' (7). While the archaeological record is too limited to prove contemporary occupation of the region, the documentary evidence that Felix, perhaps unwittingly, provides can be taken as clear indication that the region had seen human habitation prior to Guthlac's arrival if not contemporary with his occupation of the island.

A further Old English version of the Life of Guthlac, *Vercelli Homily* XXIII, follows Felix closely in calling Guthlac's hermitage an 'igland' ('island') containing 'sum mycel hlæw of eorþan geworht, þone ylcan hlæw iu geara men bræcon & dulfon for feor þingum' ('a great grave worked of earth; in days of yore men broke into and dug up that same grave for perverse things,' Scragg 383, ll. 1-4). Jane Roberts suggests that both prose translations might derive from a common source undertaken during Alfred's program of translation of important texts into Old English (1986: 376).

Treasure in a mound might be an indicator of burial: the archaeological record confirms accounts in Old English poetry of local leaders buried with weapons and armor, decorated horns and other treasures, and even the Sutton Hoo ship. Treasure might also be buried in a cave or barrow during some kind of disruption to a family or a community by, for example, famine, plague, or war. *Beowulf* contains a passage sometimes called 'the lay of the last survivor,' the words of a man leaving treasure in a barrow after the rest of his community has died. He speaks to the earth:

Heald þu nu, hruse, nu hæleð ne moston,
eorla æhte! Hwæt, hyt ær on ðe
gode begeaton. Guðdeað fornam,
feorhbealo frecne, fyra gehwylcne
leoda minra, þara ðe þis lif ofgeaf,
gesawon seledream.

> Earth, hold you now, now that heroes can no longer, what earls possessed. Behold, good people seized it from you before. War-death, savage deadly evil, has taken every one of my people; those who gave up this life saw happiness in the hall. (ll. 2247-52a)

The Britons are known to have lived in the fens, using boats much like today's punts to move about the shallow streams among the marshes and islands, fishing for sustenance and living on the islands. After the invasions of the Angles and Saxons beginning in the sixth century, Britons took refuge in the fens from the attacks of the English. Whoever buried or hid treasure on the island that Guthlac appropriated as his hermitage may have thought to return for it, or may have hoped later generations would find it. Whether those who later dug up the mound and scattered lumps of earth found treasure there is not recorded, by Felix or in other records. There is no record of who later dug a cistern in the side of the mound. But these tracks of human use demonstrate an island with a human history, despite Felix's claims that the island was an untracked wasteland.

Animals, it should be noted, also leave tracks and other traces of their habitation. The rabbits that would have lived in the fens in abundance, for instance, make herdpaths between their burrows and locations of frequent feeding; they also leave droppings as a trace of their presence. To call a territory 'trackless' ignores the evidence of animal occupation and suggests that only the tracks left by humans are worthy of notice. To call the island 'uncultivated' suggests that human use trumps animal occupation, and that animal occupation is always subject to human intervention. The description of the island is written in such a way as entirely to privilege Guthlac and his occupation over any animal inhabitant. Later in the narrative, animals as well as plants are depicted as serving Guthlac.

Guthlac determines after he has settled into his hermitage that he will no longer wear garments made of linen or wool, but only of animal skins. Linen is plant-based, while wool does not require the death of the animal for its use; but clothing made of animal skins can only be made after an animal has been killed, whether for food or specifically for its hide. There is a long tradition of hermits and saints wearing animal skins as clothing. Gale Owen-Crocker suggests that Guthlac adopted these garments in order to do penance (183), though it might have been Felix who chose the detail in an echo of earlier saints' lives. In any case, Felix uses it to establish something about Guthlac, as the human subject of the *Vita*, and not because of any attention to the life of the animal killed for its skin. As discussed in the following chapter, animals in some of the *Exeter Book* riddles are given voices that protest their killing for human utility. But in Felix's *Vita* there is no suggestion that the animal possesses agency or is worthy of moral consideration.

Animals and reeds, and even water, exist within Felix's *Vita* to serve Guthlac and demonstrate his saintly connection to God. Felix tells a story of a visitor who finished writing a document on parchment and walked away, leaving it lying where a crow is able to pick it up and carry it away. Guthlac gets his boat and pursues the crow in order to retrieve the parchment.

> Dein, cum ad aliquod stagnum haud procul a praefata insula situm devenisset, conspicit non longe in media planitie stagni unam harundinem curvato cacumine stantem, quae stagni tremulis quassabatur undique limphis; in cuius fastigio aequiperatas, scedulas aequali lance pendentes, velut ab humana manu positas, cerneres. Mirabile dictu! tangi, non tactae, contiguis videbantur ab undis. At ille frater arripiens de harundine cartam, cum magna admiratione grates Deo persolvens, venerantiam validae fidei de eo quod contigit venerabili viro Dei Guthlaco conferens, unde egressus domum reversus est.
> Then when he had reached a certain pool not far from the same island he saw nearby, in the middle of the pool, a reed standing with its top bent down and shaken on every side by the moving waters of the pond; on the very top could be seen the very leaves of parchment hanging exactly balanced as though they had been placed there by a human hand, and, marvellous to relate, they were apparently being touched by the waves around them and yet were intact. And the brother, snatching the document from the reed, gave thanks to God in much amazement, at the same time showing great respect for the steadfast faith of the venerable man of God Guthlac with regard to what had happened; and so departing thence he returned to his dwelling. (117, 118)

The Old English version departs from the Latin text slightly in that the visitor, rather than Guthlac, rows to the point where he finds the leaf of parchment:

> Mid þy he þurh þa fenland reow, þa com he to sumum mere þe wel neah þæt egland wæs: þa wæs þær on middan þam mere sum hreod-bed; þa hangode seo carte on þam hreode efne swa hig monnes hand þær ahengce: and he sona þa bliþe feng to þære cartan, and he wundriende to þam Godes were brohte: and he þa se eadiga wer Guthlac sæde þæt þæt nære his geearnung ac Godes mildheortnys.

> When he had rowed through the fenland, then he came to a mere that was near the island. In the middle of that mere there was a bed of reeds, and there the charter hung on a reed even as if a man's hand had hung it there: and immediately he happily seized the charter, and wondering, brought it to the man of God. And he, the blessed man Guthlac, said that that was not (because of) his merit, but God's mercy. (50)

The reed and the water are deployed in this passage to demonstrate Guthlac's sainthood and connection to God. The reed of Guthlac's fen is described purely in terms of its ability to demonstrate Guthlac's saintliness.

Ecocritics use the term 'instrumental' to refer to a human-centered view in which natural phenomena are important only insofar as they are useful to humans in real life or have metaphorical force in literary works. The reed near Guthlac's hermitage functions instrumentally in that it is of interest only in its utility to Guthlac, in literal terms, in saving the material document that his visitor was writing, and in spiritual terms, in demonstrating Guthlac's superiority such that the plants bend themselves to his will. As described in Chapter 2, Beowulf takes a similarly instrumental view of the sea-creatures he slaughters while swimming with Breca.

In contrast, the Reed of *Exeter Book*, Riddle 60 speaks in its own voice and asks the reader to guess what creature it is, saying 'Ic wæs be sonde, sæwealle neah, / æt merefaroþe, minum gewunade / frumstaþole fæst' ('I was by the sand, near the sea-wall, on the waves, my dwelling a fixed original abode,' ll. 1-3). The Reed of Riddle 60 is made into a pipe, used to make music or send messages between men, so it is turned into something of utility to humans, but the start of the description locates it in its original dwelling, growing on its own terms. The riddle opens the possibility of a point of view that allows agency to the plant independent of human concerns. The human-centered view of the non-human world depicted in the various versions of the *Guthlac* narrative, that places other humans as well as plants and animals on a similar plane of utility to the saint, is counter-balanced in the *Exeter Book* riddles, demonstrating that a non-instrumentalizing perspective was a possibility for Anglo-Saxon writers and scribes. In the Guthlac narratives, however, everything is put to the service of a Christian narrative that elevates spiritual expression and the saintly behavior of the male hermit above all other considerations.

Felix gives further examples that depict the natural environment used instrumentally to demonstrate Guthlac's piety. Two crows are 'infesta,' ('hostile, mischievous,' 118) and make a habit of damaging and destroying Guthlac's possessions, but he reacts only with patience even though the birds are 'velut inprobi praedones rapiebant' ('like shameless robbers,' 120). Felix depicts the crows in anthropomorphized terms, not by way of understanding them or their actions but simply in order to make a point about Guthlac's character. The passage might recall Guthlac's predation on the bodies of enemy soldiers, since crows, as members of an Anglo-Saxon audience would well have known, were carrion-eaters; the point seems to be to emphasize Guthlac's transformation in his devotion to God. Felix adds:

> Erga enim omnia eximiae caritatis ipsius gratia abundabat, in tantum ut incultae solitudinis volucres et vagabundi coenosae paludis pisces ad vocem ipsius veluti ad pastorem ocius natantes volantesque subvenirent; de manu enim illius victum, prout uniuscuiusque natura indigebat, vesci solebant. Non solum vero terrae aerisqac animalia illius iussionibus obtemperabant, immo etiam aqua aerque ipsi veri Dei vero famulo oboediebant.
> For the grace of his excellent charity abounded to all creatures, so that even the birds of the untamed wilderness and the wandering fishes of the muddy marshes would come flying or swimming swiftly to his call as if to a shepherd; and they were even accustomed to take from his hand such food as the nature of each demanded. Not only indeed did the creatures of the earth and sky obey his commands, but also even the very water and the air obeyed the true servant of the true God. (120, 121)

Although this seems to contradict the earlier narrative of hostile and predatory crows, Felix states outright that the non-human environment – encompassing animals as well as water and air – serves Guthlac as a demonstration of his devotion to God. The Old English abbreviates the description of the wicked crows as well as the passage that follows:

> na læs þæt an þæt him þa fugelas underþeodde wæron, ac eac swa þa fixas, and wilde deor þæs westenes ealle hi him hyrdon, and he hym dæghwamlice andlyfene sealde of his agenre handa, swa heora gecynde wæs.
> No less the one that the birds were subordinate to him, but also the same the fish, and wild animals of the wilderness all obeyed him, and he gave them food every day from his own hands, according to their kind.

The passage condenses Felix's longer discussion of Guthlac's dominion over the earth. According to Felix, earth and sky, water and air, birds and fish obey the commands of Guthlac. The Old English translator eliminates earth and sky from the entities that obey Guthlac, but emphasizes that the animals that Guthlac feeds are not domesticated but wild, a detail absent from the corresponding Latin passage. The fact that the Old English translation elides the intractable crows eliminates an apparent contradiction in the Latin *Vita* to make Guthlac's dominion over the wild animals even more complete.

Both Felix and the translator of the Old English version use landscape alongside animals and material objects to demonstrate Guthlac's saintliness. They do not suggest that Guthlac should be in any way accountable to the environment around him, whether the earth in which he builds his hermitage, the crows that harass him, or the reeds that hold up the document his visitor has been writing. Moreover, the landscape is static in Guthlac's occupation. Felix states that Guthlac first moved to his refuge on St. Bartholomew's day, 'aestivus temporibus' ('in the summer time,' 88, 89), and that Guthlac took ill shortly before Easter after several years have gone by. But there is no depiction of changing seasons or of any other shifts in the landscape that Guthlac occupies. In terms of Buell's early articulation of what kinds of texts fit a canon of environmental literature, the *Vita Guthlaci* does not belong. Its failure to show landscape as changing in any way demonstrates that texts that avoid presenting the environment as existing in any fashion independent of human concerns are not recent innovations, and that it is possible for Anglo-Saxon authors and audiences to 'read' landscape in this way, despite the fact that they live in what later periods have romanticized as a pastoral environment.

Britons as/and Demons

When he takes up residence 'inter nubilosos remotioris heremi' ('amid the gloomy thickets of that remote desert,' 90, 91), Guthlac arms himself once again, in spiritual terms that echo his actual taking up of arms as a youth:

> Deinde praecinctus spiritualis armis adversus teterrimi hostis insidias scutum fidei, loricam spei, galeam castitatis, arcum patientiae, saggitas psalmodiae, sese in aciem firmans, arripuit.
> Then, girding himself with spiritual arms against the wiles of the foul foe, he took the shield of faith, the breastplate of hope, the helmet of chastity, the bow of patience, the arrows of psalmody, making himself strong for the fight. (90, 91)

Guthlac is repeatedly attacked by demons in various forms, in echoes of passages adapted from other saints' lives as well as of Jesus' battle with demons in the wilderness. Three of these demonic descriptions are of particular interest from a post-colonial, ecocritical perspective. In one of the earlier attacks, demons that Felix characterizes as 'teterrimus inmundorum spirituum catervis' ('horrible troops of foul spirits,' 100-03) fill Guthlac's hermitage. They have many human characteristics but are deformed or monstrous in form:

> Erant enim aspect truces, forma terribiles, capitibus magnis, collis longis, macilenta facie, lurido vultu, squalida barba, auribus hispidis, fronte torva, trucibus oculis, ore foetido, dentibus equineis, gutture flammivomo, faucibus tortis labro lato, vocibus horrisonis, comis obustis, buccula crassa, pectore arduo, femoribus scabris, genibus nodatis, cruribus uncis, talo tumido plantis aversis ore patulo, clamoribus raucisonis.
> For they were ferocious in appearance, terrible in shape with great heads, long necks, thin faces, yellow complexions, filthy beards, shaggy ears, wild foreheads, fierce eyes, foul mouths, horses' teeth, throats vomiting flames, twisted jaws, thick lips, strident voices, singed hair, fat cheeks, pigeon breasts, scabby thighs, knotty knees, crooked legs, swollen ankles, splay feet, spreading mouths, raucous cries. (100, 102)

The passage follows a typical medieval sequence in describing a human figure by starting with the head and working toward the feet, though the details of 'spreading mouths [and] raucous cries' are out of sequence. Aside from 'horses' teeth' and 'pigeon breasts,' these demons have entirely human-like features, though described in terms of filth, illness, and deformity. The Old English translation follows the Latin closely, though the Anglo-Saxon translator uses doubled terms of opprobrium for emphasis, for example rendering the Latin 'squalida barba' ('filthy beards, shaggy ears') as 'fulice and orfyrme on heora beardum' ('foul and squalid in their beards,' 34).

Interestingly, the Old English translator renders 'gutture flammivomo' ('throats vomiting flames') as 'him wæron þa þrotan mid lege gefylde.' Rather than the more unambiguous word 'fyr' ('fire'), the translator (or possibly scribe) here uses a word that can mean either 'flames' or 'lies,' so that the demons could be understood as breathing fire, and thus monstrous in the way of the dragon in *Beowulf* – or, more mundanely, as having 'throats filled with lies.' The latter reading allows an interpretation of these demons as more human, a suggestion echoed by a passage three chapters later in which the demons much more clearly resemble humans.

Felix writes that in the time of King Coenred, the Britons, whom he calls 'infesti hostes Saxonici generis' ('implacable enemies of the Saxon race'), had been attacking the Mercians. The Saxons had been the invaders, in a war in which Guthlac had been an apparently enthusiastic participant, yet in casting the Britons as 'infesti hostes,' Felix implies that they had been the aggressors and that the Mercians had been justified in slaughtering them. In using the same adjective for both Britons and crows, Felix blurs the distinction between human and animal – to the detriment of Britons, but not of Anglo-Saxons. Guthlac hears a crowd approach his hermitage and recognizes them as Britons based on their 'strimulentas loquelas' ('sibilant speech,' 108-11). He prays, uttering the beginning of Psalm 68 (67), 'velut prophetico' ('as if prophetically,' 110, 111). Felix does not include the text of the psalm; it begins 'Exsurgat Deus, et dissipentur inimici ejus' ('Let God arise, and let His enemies be destroyed,' Psalm 68:1). The Old English version is quite similar; it omits reference to the 'sibilant' quality of the demons' speech, but refers to them specifically as 'British':

> Guðlac ... gehyrde he mycel werod þara awyrgedra gasta on bryttisc sprecende, and he oncneow and ongeat heora gereorda for þam he ær hwilon mid him wæs on wrace.
> Guthlac ... heard a great troop of the cursed troop speaking in British, and he knew and understood their language because earlier he had been among them for a while as an exile. (42)

The earlier section describes the demons as humanoid, yet is ambiguous about their actual humanity, allowing an interpretation of monstrosity. This passage clearly identifies the demons harassing Guthlac with human opponents of the Mercians.

Two chapters later, Felix, followed by the Old English translator, describes demons in the form of a herd of wild animals. Felix characterizes them as 'variorum monstrorum' ('various monsters,' 114, 115), while the Old English translator changes this to 'wildeora and wurma' ('wild animals and dragons,' 46). Guthlac is threatened by a lion with 'dentibus sanguineis' ('bloody teeth,' 114, 115) rendered in old English as 'blodigum tuxum' ('bloody tusks,' Goodwin 48) as well as by a bull, a bear, wolves and ravens, and snakes. Felix's use of 'monstrorum' might, then, be better translated as 'portents' or even 'unusual things' – as bears and bulls and lions would indeed be in the East Anglian fenlands.

The identification of demons as Britons – or of Britons as demons (?) (O'Brien O'Keeffe 2003: 21) recalls Guthlac's earlier life as a warrior; a further echo is generated by the comment that he recognizes their speech from his sojourn among them, which must have been during his war years, because he went from childhood to being a warrior to being a monk. As O'Brien O'Keeffe writes, 'While it is perfectly clear from both the chapter heading and the conclusion of this chapter that Guthlac's persecutors here are demons, not men, they are presented throughout the narrative as a real, encroaching army, closing in through the swamp, burning and killing' (2003: 21).

The scenes in which the hermitage is imaged as occupied by demons echo the scenes of battle earlier in the narrative in the ways that they construct Guthlac as righteous in his aggressions against his neighbors. Guthlac slaughters Britons in the quest for political control over Mercia. He encounters the traces of inhabitants who might have been Britons when he first approaches the 'untracked' island on which he will make his hermitage. After he has retreated there alone, he has visions of demons with the voices of Britons. As O'Brien O'Keeffe has argued, 'Felix's *Vita S. Guthlaci* is first and foremost a discourse of contested territory fundamentally altered through acquisition, purgation, and habitation.... Through Penwalh, and indeed Guthlac himself, Felix firmly imbricates his narrative in the geographic and political struggles in Southumbria of the later seventh century' (2001: 3-4). The various versions of the *Life* attempt to cast Guthlac's occupation as a religious retreat to a desert, but details that emphasize colonizing political contexts of the period continually reassert themselves. Conflation of the literal battles of Guthlac's warrior years with the spiritual battles of his time in the hermitage has the effect of legitimating the territorial battles of Mercian, Anglo-Saxon invaders against the native Britons who, in Felix's narrative, are the ones 'harassing' the Anglo-Saxons. Casting Guthlac's battle for possession of his fenland hermitage in terms of battles against hordes of demons rather than against human inhabitants of the place normalizes the subjugation of others, because demons are by definition evil. A hermit's victory over the temptations and tortures of demons becomes evidence of devotion to God, in a paradigm that obscures any possibility of negative interpretation. The fenland possession battle is juxtaposed, in Felix's narrative, with the battle against the Britons, earlier inhabitants of the land that the Angles and Saxons are in the process of colonizing. As O'Brien O'Keeffe further argues, 'Felix's particular deformations of the discourse of the desert produces the fen as a theatre of political anxiety, as a place to stage the meeting of unimaginable difference and to produce the result in the comforting choreography of Christian triumphalism' (2001: 25). The identification of demons with Britons collapses distinctions between Guthlac's political battles as the invader and his spiritual contest with the demons he, or Felix, imagines as his tempters and opponents. As Siewers argues, 'such issues of ethnicity are relevant to descriptions of the Fens in Felix as a last refuge for the demonic spirits he associates with the Britons, as well as to the description of the landscape of the Fens themselves' (2003: 12). The religious narrative of Guthlac's withdrawal as a hermit conceals colonizing ethnic politics.

The narrative about Guthlac's move to the hermitage simultaneously normalizes human dominion and power over animals and the non-sentient world, and depicts it as a Christian value. Guthlac's power over non-human creatures extends to the small wild animals native to the fenlands as well as over the demons that are depicted as 'unnatural,' aggressive beasts. Animals are cast as demonic others or as servants of the devout Guthlac, in either case depicted as dichotomously opposed to rather than aligned with humans on any kind of a continuum. The intertwining of human, monstrous and animal in depictions of the demonic, all opposed to Guthlac's privileged position in gender, class, and ethnic hierarchies, functions to establish a small class of people – 'englisc,' Christian, male, upper-class – in opposition to all others, with power over land, animals, and other people analogously legitimized through the idea of devotion to God as the highest of values.

Guthlac A and the 'beorg'

In addition to the Old English prose translation discussed above, Felix's *Vita Guthlaci* was adapted into two Old English poems, preserved as the second and third works in the poetic compilation known as the *Exeter Book*, which also includes the riddles discussed in the following two chapters. The second of the poems, known as *Guthlac B*, concerns Guthlac's death and ascendance to heaven, and draws more closely on Felix's *Vita* than does the first, *Guthlac A* (Roberts 1988: 2). But it does not include the sections of the *Vita* that concern Guthlac's retreat to the hermitage and Felix's depictions of the landscape he settles , or his interactions there with various other creatures.

Guthlac A, in contrast, contains passages corresponding to those in Felix's *Vita* and in the prose translation concerning Guthlac's occupation of the island and the attacks from demons. Following a description of heaven and an exhortation to readers to have faith so that they may attain it, *Guthlac A* begins the narrative of the saint at the point when Guthlac decides to become a hermit, eliminating the account of his early life and of his years as a warrior. Many details of Guthlac's interactions with the environment and the demons that harass him are changed, but the sense of Guthlac as conqueror over the landscape and all of its occupants remains.

Defining the 'beorg' that Guthlac occupies in the poetic text has occupied much scholarly attention, with scholars arguing for either hill or grave mound. The 'tumulus' or 'hlaw' ('burial mound') of the *Vita* and the Old English prose translation is rendered in the poetic account as 'beorgseþel' (l. 102) and 'beorg' (ll. 148, 193, 209, etc.). Both Bosworth-Toller and J. R. Clark Hall translate 'beorgseþel' as 'mountain dwelling,' taking 'beorg' to mean 'mountain.' 'Beorg' can also mean 'barrow,' and the *Vita* and the Old English translation clearly refer to a burial mound on a low-lying island in the fens, but other passages in *Guthlac A* appear to describe Guthlac's hermitage as situated on a hill. The poem states that Guthlac became an example to many others in Britain 'siþþan biorg gestah' ('after he climbed the mountain,' l. 175b). Later in the poem the line is echoed when Guthlac tells the demons 'nu ic þis lond gestag' ('now I climbed this land,' l. 307b) and the narrator adds, 'he eft gestag / beorg on bearwe' ('he again climbed his hill in a grove,' ll. 428b-29a). The combination of 'biorg' with 'gestigan,' meaning 'to climb' or 'to descend' suggests that Guthlac's hermitage is on a hill or mountain rather than in a low-lying fenland grave mound. Beyond these references to climbing, however, there is little textual evidence within the poem itself to aid in interpretation.

Reichardt takes the 'beorg' to be a mountain and a 'symbol of interior spiritual achievement' (335); while Wentersdorf (1978) argues that 'beorg' is a direct translation of Felix's 'tumulus' and Shook takes the 'beorg' to be a fenland burial mound, 'perhaps of no great antiquity in Guthlac's time' (4). Jane Roberts points to the paucity of textual evidence within *Guthlac A*, arguing that 'there is not even any sufficiently explicit detail to allow firm identification of the *beorg* as "gravemound." ... Overall the landscape of Guthlac A lacks sharp definition' (1988: 11). Reading *Guthlac A* in the context of Irish and Welsh texts, Siewers notes, like Roberts, the lack of specificity concerning the 'beorg,' and compares the hermitage of *Guthlac A* to Grendel's mere in *Beowulf* (2003: 2). He argues for a polysemous reading of the 'beorg:' 'The appositive meanings of *beorg*, pagan and Christian, were probably part of the intent of a poet whose theme rode a *mearclond* between different British landscapes' (Siewers 2003: 24). Clarke suggests that the passage casts the landscape as '*locus amoenus* or conventionally delightful landscape' (36) and 'a place of dual potential, for suffering and testing' (48), and argues that the exact meaning of 'beorg' is unimportant (47). Most recently, Johnson argues that Guthlac's hermitage undergoes a transformation within the poetic narrative from mountain to a 'plain of victory' ('sigewong,' l. 742) in an echo of Isaiah 40: 4: 'Every valley shall be lifted up, and every mountain and hill be made low' (309-10).

Like the Latin *Vita* and the Old English translation, the poem depicts Guthlac's hermitage as a desert or wilderness:

> Wæs seo londes stow
> bimiþen fore monnum, oþþæt meotud onwrah
> beorg on bearwe, þa se bytla cwom
> se þær haligne ham arærde
>
> The place in the land was hidden from men until God uncovered a mound/hill in a wood, where the builder came, who there for the holy one raised a home. (ll. 146b-49)

As noted above, the prose versions similarly state that the region was unknown, but then introduce a man named Tatwine who showed it to Guthlac. The poem eliminates this contradiction by saying that God revealed the spot to Guthlac alone. The hermitage is 'on westenne' ('in the desert' or 'in the wilderness,' l. 209), and the wilderness is further characterized as vast ('wid is þes westen,' l. 296a). The region is wooded, like its counterpart in the *Vita*, but the poem nowhere refers to it as an island.

Genesis A similarly modifies a textually transmitted description of wilderness. In the biblical Genesis, God tells Abraham to take Isaac to a mountain ('montium,' Gen. 22: 2) in Moriah to slaughter him. In the Vulgate, there is no further description of the place. The Old English *Genesis A*, however, expands upon the text. The 'mountain' of the Latin becomes a 'steape dune' ('high mountain,' l. 2854) on the very edge of the highlands ('hrincg þæs hean landes,' l. 2855). Abraham and Isaac travel 'wegas ofer westen' ('paths over the wilderness,' l. 2875) that lead to 'steape dune... þæt he on hrofe gestod hean landes' ('a high mountain... so that he stood on the summit of the high land,' ll. 2897b, 2899). *Genesis A* adapts and elaborates on the Vulgate's 'montium,' emphasizing the terrain's remoteness and steep highlands.

This might reflect topographical reality as recounted by Anglo-Saxon pilgrims, traders or other travelers who had been to Rome across the Alps. Such a crossing, whether on foot or on the back of a mule or horse, would require crossing difficult terrain far from settlements of any size. The only surviving record of a journey between England and Rome during the Anglo-Saxon period is that recording the return itinerary of the Archbishop Sigeric in 990, which records that he traveled from Italy over the St. Bernard

Pass at 8110 feet into the Rhone Valley of what is now Switzerland and thence across France, eventually arriving at the English Channel (Ortenburg passim). The Romans had constructed a road through the pass by 47 CE. Saint Bernard built a hospice in the pass to help travelers, but not until the eleventh century, though there were likely earlier travelers' refuges, if not in the pass itself then on the approaches to it from either side.

Beowulf is located in Denmark, not England, and so in geographically remote terrain; there is no prose source or analogue with which to compare the poetic treatment of the landscape. But to an even greater extent than England, Denmark is flat country; with a high point of 568 feet and numerous islands, it resembles, as Siewers has pointed out, Guthlac's fenland. Yet at some point in the transmission of the narratives about Beowulf and his foes, the landscape Grendel and his mother inhabit is re-imagined as a deserted, craggy wasteland. The pool through which their underground lair is accessed lies in terrain characterized by steep cliffs and sharp drops. Hrothgar tells Beowulf that Grendel and his mother occupy 'wulfhleoþu, windige næssas' ('wolf-hills, windy headlands,' l. 1358); a stream descends 'under næssa genipu' ('under the shadow of the cliff,' l. 1360). The transformation of remote landscape into wilderness highland in the adaptation of both *Genesis A* and *Guthlac A* as well as in *Beowulf* suggests that Anglo-Saxon authors and audiences imagine 'wilderness' as mountainous rather than marshy terrain. A burial mound can be climbed, but the references to ascending the 'beorg' suggest a hill of some size. The absence of any synonym in the poem for 'hill' or 'mountain,' however, (in contrast, for instance, to *Christ and Satan*, which includes references to 'hyll' and 'dun,' or *Genesis A*, which refers to 'dun,' 'heah land,' and 'hrofe [roof]') leaves the 'beorg' of *Guthlac A* open to both possible interpretations, as both barrow and mountain simultaneously.

Much as the prose versions deny that the hermitage is inhabited, but describe three previous layers of human use, the poem likewise suggests prior occupation of the hermitage by some 'feond' ('devil, enemy,' l. 136) who 'þær ær fela / setla gesæton' ('had made many settlements there earlier,' ll. 143b-44a). The poem then states, 'oft þær broga cwom / egeslic ond uncuð' ('terror often came there, fearsome and unknown,' ll. 140b-41a). Whatever 'feond' had previously occupied the place, whether human enemy or non-human demon, it is terrifying precisely because it is unknown. The demons, 'teonsmiðas tornes fulle' ('evil-doers, full of fury,' l. 205) are the other, much as the Britons are characterized as 'other' by the colonizing Saxons; characterizing demons as unknown contributes to characterizing unfamiliar human populations as demonic. The demons accuse Guthlac of occupying their refuge 'for wlence' ('because of pride,' l. 208). Demons are, categorically, liars, whose goal is to trick Guthlac into giving up his devotion to God. Thus the claim that they are the rightful occupants of the land attains the predetermined status of outright falsehood. Yet the claim that the space is unoccupied and unknown, in conjunction with the acknowledgment that it was known to and occupied by the demons, recuperates the contradictions of tracked wilderness found in the Latin *Vita* and the Old English prose translation. Another trace of prior occupation for the 'beorg' comes in the narrator's comment that Guthlac's choice of that place for his hermitage was 'nales þy he giemde þurh gitsunga / lænes lifwelan' ('not at all because he cared, out of avarice, to obtain the riches of this world,' ll. 150-51a). This line points to the possibility that the 'beorg' is not a hill, but a grave after all, filled with grave-goods that Guthlac, the poem thus takes pains to point out, is not at all interested in obtaining.

Siewers argues that the text, following patristic thought, presents both the demonic previous occupants and nature itself as problematic:

> Somewhat paradoxically, Augustinian theological emphases on the corruption of nature, extended to natural landscape and its ancestral associations with indigenous culture, empowered the Anglo-Saxon ideological project of superimposing a new cultural landscape on Britain's most fertile land areas, in narrative landscapes based on a sense of Anglo-Saxon culture as God-chosen and hegemonic that erased textually the presence of earlier inhabitants as thoroughly as Old English linguistically replaced Romano-Celtic languages in those areas. The presence of indigenous Romano-Celtic linguistic cultures that were Christianized long before those of Anglo-Saxon realms, and which exerted a large continuing influence on the latter, was thus conveniently erased or subsumed (2003: 3).

Felix and his translators and adaptors into Old English prose and poetry drew on a long Christian Latin cultural tradition to enable colonial expansion, but also fused it with Germanic habits of invasion and plunder and forged a new synthesis that justified the Anglo-Saxons as the rightful possessors of territory formerly held by Britons using Christian faith as the measure of suitability.

Conclusion

The refuge to which Guthlac retreats to seek solitude is transformed between Felix's *Vita* and *Guthlac A* from the East Anglian fenland to an apparently mountainous desert consistent with other Old English literary depictions of wilderness. But the assumptions that govern Guthlac's appropriation of the place remain constant across the Latin and Old English texts. While the space has clearly seen previous occupation, calling it 'uncultivated' makes a cultural claim asserting the superiority simultaneously of Mercians over Britons and Saint Guthlac over pagans. As Siewers argues, Augustinian theology is behind Guthlac's colonizing of the fen, but Felix merged Christian textual heritage with Anglo-Saxon warrior culture. 'In the textualizing, and resulting distancing, of the natural and the spiritual, Augustine's writings helped shape the theme of environmental utilitarianism so important in Western culture, which emerged early in Anglo-Saxon literature for reasons of political ideology' (2003: 8).

In *Beowulf*, Hrothgar's queen, Wealhtheow, is sometimes read as a Welsh queen who has married him in an attempt to establish peace between their communities, suggesting the possibility of inter-group ties. But in the *Vita Guthlaci*, Felix depicts groups of people other than those of Guthlac's 'tribe' as completely insignificant beneath Guthlac's masculine and Christian concerns, first to demonstrate his martial prowess and then to show his saintly nature in the hermitage. The ideas that enabled empirical expansion, the subjugation of people outside of Europe, the destruction of land and the unrestrained slaughter of animals, are deeply embedded in English culture, preceding actual colonization outside of Europe by several centuries, and enabling and justifying later colonial aggressions. The idea that land, animals, and other humans are available for English conquest is not a side effect or a 'bug' of territorial expansion, but a precondition for that expansion. Huggan and Tiffin describe their own hopes for their study of post-colonial ecocriticism:

> In reaching out across languages and cultures, postcolonial ecocriticism is paradoxically driven – as is this book – by the impossibility of its own utopian ambitions: to make exploitation and discrimination of all kinds, both human and nonhuman, visible in the world; and, in so doing, to help make them obsolete. (Huggan-Tiffin 16)

The imagined wildernesses of Anglo-Saxon texts set the stage for colonial expansion. Imagining these 'wild' lands as more thoroughly distant from actual Anglo-Saxon experience aids in conceiving of them as impervious to cultivation and therefore devoid of human habitation. But even viewed from geographically and temporally closer standpoints, as in Felix's *Vita* written soon after Guthlac's death, land and its occupants are presented as justifiably occupied in the twin visions of Christian piety and Anglo-Saxon military goals which, combines, are deployed to legitimate foreign occupation and territorial expansion. The oral, documentary, and literary traditions that informed the various versions of Guthlac's life allow for constructions of population and environment that suggest that it is normal for lands, animals, and peoples to be subjugated for utilitarian purposes.

The two chapters that follow move from epic and biblical poetry to the *Exeter Book* riddles. The paradigms of human domination that suffuse *Andreas*, *Beowulf*, *Exodus*, *Genesis*, and *Guthlac* are reversed, if only briefly, to open doors to a different possible construction of the environment and human interactions with it. The riddles take on the voices of animals, trees, and ore from the earth that are transformed into objects used by humans; humans are described as the enemy and are othered by the texts of the *Exeter Book* riddles. In Chapter 6, I extend Timothy Morton's notion of hyperobject, which he coined as a way of thinking about climate change, to social constructions, examining how objects are metaphorically cast as humans and humans reduced to objects and revisiting the ecofeminist philosophies explored in Chapter 2. I explore the intersections among ecofeminism, postcolonial ecocriticism, and object-centered philosophies to show how the literature of the Anglo-Saxons anticipates and even directly articulates hierarchical and hegemonic positions that still persist in contemporary European and North American cultural formations about the priority of some humans over others.

5 Animal Natures

Introduction

In the preceding chapters I have considered to some extent the place of animals in *Beowulf* and the Guthlac narratives, and their functions with respect to the humans in the texts. The so-called 'sea-beasts' in the mere seem to enhance Grendel's mother's monstrosity, while Beowulf's ability to kill while swimming serves to demonstrate his power and masculinity. In this chapter on 'animal natures' I dig deeper into Anglo-Saxon representations of animals, focusing on several of the *Exeter Book* riddles in the context of the formulaic 'beasts of battle' scenes in *Elene*, *Exodus*, and *Genesis A*.

Animal studies does not always overlap with broader environmental concerns. Environmentalists are concerned with issues of extinction and how animal movements and extinctions reflect climate change and other global issues, while scholars and activists concerned with animals focus on philosophical issues of rights and agency, as well as how the characteristics of animals define or limit the human (or not). Both approaches are valid, indeed necessary. This chapter focuses on how depictions of animals and humans interact in medieval texts, and attempts to locate those interactions within broader environmental questions.

The meaning of the word 'animal' is difficult to pin down in either modern or medieval contexts. As Susan Crane notes, the word 'resists definition. *Animal*, synonymous with *beast* in Middle English, sometimes encompasses and other times contrasts with what is meant by *human*' (1). Karl Steel, however, argues that in the medieval period, what it means to be human is defined in relation and in opposition to the animal. In contemporary theoretical formulations, as Sarah Kay points out, 'it is common to distinguish two directions in animal studies: one, animal activism and the promotion of animal rights, the other, posthumanism in the sense of reconsidering human specificity in relation to (other) animals.' She notes, however, that 'this distinction between advocacy and ontology is far from robust' (Kay 14). This chapter focuses neither on animal rights nor on defining humanity in contrast to or in continuity with animals. Instead, I seek to clarify the ways in which Anglo-Saxon literary texts, including several of the *Exeter Book* riddles as well as *Beowulf* and a few other epic poems discussed in less detail present conflicting ideas about animals and human relationships with them. In Chapter 2, I examined the unthinking slaughter of sea creatures and the construction of the monstrous and dangerous by way of justification for their death. Here, I review formulaic language about 'beasts of battle' in contrast to the non-instrumentalizing descriptions of birds in the *Exeter Book* riddles.

There are many environmentally focused ways of thinking about animals and attempting to theorize why they should matter. They might be assumed to matter because of their impact on humans, or simply because they exist and are therefore taken to be of importance in their own right, but the important thing is that they matter. Ecocriticism and animal philosophies are often concerned primarily with present conditions, but as Bruce Holsinger has written: 'medieval culture has much to teach the wider fields of animal studies and animal ethics about some of the fundamental questions that have defined this domain of inquiry' (617). As chapters on wilderness, ruins, the sea, and objects also demonstrate, the past is crucial to our understanding of the present.

In the *Exeter Book* riddles, the depictions of animals are complex and unstable. Considering the riddles in comparison or contrast to other texts from the period further complicates the issues, in that animal descriptions in the riddles are very different from those in epic poems, saints' lives, and biblical adaptations. Numerous animals are described in the riddles in some detail; many are narrated in the voice of the animal itself, sometimes even in protest against the cruelty of the human 'enemy.'

Over the course of several days in 1997 Derrida gave a series of lectures in which he contemplated animals – the gaze of his cat, and the tendency of philosophers to treat all animals in the singular in drawing a line between 'the human' and 'the animal' as one of the bases of defining the human.

> Beyond the edge of the so-called human, beyond it but by no means on a single opposing side, rather than 'the Animal' or 'Animal Life,' there is already a heterogeneous multiplicity of the living, or more precisely... a multiplicity of organizations of relations between living and dead.... These relations are at once close and abyssal, and they can never be totally objectified. They do not leave room for any simple exteriority of one term with respect to another. It follows from that that one will never have the right to take animals to be the species of a kind that would be named the Animal, or animal in general (2002: 399).

He further declares that anyone who uses the term 'the Animal' as if it refers to a not-human totality 'utters an asinanity' and 'confirm[s]... his complicit, continued and organized involvement in a veritable war of the species' (2002: 400). But while Derrida insists on multiplicity when it comes to characterizing animals, he still draws a clear line between animals and humans, one that many animal studies scholars would reject.

Many Anglo-Saxon texts articulate or imply such a totalizing vision of 'the animal' in contrast to and perhaps in construction of the human. One example is Beowulf's treatment of the 'sea beasts' he slaughters during his swimming contest with his boyhood friend Breca, discussed in Chapter 2. The *Exeter Book* riddles, however, depict animals in a much more individualistic manner, although in complicated and sometimes contradictory relationships to humans. This chapter begins by considering how Anglo-Saxon texts normalize the eating of animals, and then discusses how the riddles depict killing animals as complex, with humans called 'the enemy.' It then moves to a discussion of animals represented in their native habitats and described in distinct detail, with the wild bird riddles in contrast to the formulaic language of the 'beasts of battle trope.' Taken collectively, these riddles show the instability of the categories of human and animal. The following chapter continues a discussion of the riddles concerned with objects and examines the shifting and unstable boundaries between objects, animals, and humans, particularly those at the margins of cultural power matrices.

Eating Animals As Cultural Norm

While the animal rights philosopher Peter Singer argued in *Animal Liberation* that humans should not eat animals, much philosophical and literary analysis avoids the issue of human consumption of animals, focusing instead on the interrelationships between humans and animals and how representations of each construct our understandings of the other. That the Anglo-Saxons ate meat of various kinds is clear from the documentary and archaeological record. The aristocracy ate meat in higher quantities and greater variety than the peasantry; the sixth-century Benedictine Rule forbade monks from eating 'carnium vero quadrupedum omnimodo' ('flesh from four-footed animals' (ch. 39, § 71), though this still leaves birds and fish as unquestionably available for human consumption.

Ælfric's *Colloquy* lists hunters and fishermen among the professions. The Master in the *Colloquy* asks one of the pupils what he has eaten today, and he replies that he still eats flesh, because he is a child, in addition to 'Wyrta & æigra, fisc & cyse, buteran & beana & ealle clæne þingc ic ete mid micelre þancunge' ('Vegetables and eggs, fish and cheese, butter and beans and all clean things I ate, with great thankfulness,' ll. 288-89). According to Felix's

Vita, one signal of Guthlac's saintliness is his ascetic attitude toward food: he eats once a day, after sunset, and then only 'ordeacei panis particula et lutulenae aquae poculamento' ('a scrap of barley bread and a small cup of muddy water,' Colgrave 94-95). Avoiding animal flesh as food is not, for Guthlac, undertaken in deference to any sense of the right of an animal not to be used in this way, but as an expression of his own self-denial and spirituality.

Sarah Salih notes that in contemporary culture, the eating of animals appears to be normalized to the extent that a vast majority of people never think about it (61). In Old English poetry, eating meat is a norm to the extent that it is almost never mentioned, unless to point out that someone like Saint Guthlac ate only bread – but it is the consumption of bread that is explicitly referenced, rather than abstention from animal foods. While descriptions of feasting are common enough in Old English poetry to warrant designation as a 'type-scene,' in the actual feast scenes, there is little discussion of eating and much more of drinking. The feast scenes in *Beowulf* describe Wealhtheow handing a goblet of wine to the various warriors in turn, but do not comment on what the warriors are eating. *Judith* likewise gives attention to Holofernes' excessive drinking, but makes almost no mention of food, even though the biblical source for the poem lays emphasis on her bag of cheeses. Riddles 11, 27, and 28 refer to the dangers of excess when consuming mead, wine, or ale, while Riddle 63 describes a cup that could be used for drinking wine. Though there are references to plants and animals commonly eaten – hens, onions, and garlic – only one of the riddles describes a person eating an animal for food.

Riddle 77 is written from the perspective of an oyster whose peaceful life in the sea is interrupted by the incursion of a man: 'Nu wile monna sum / min flæsc fretan, felles ne recceð, / siþþan he me of sidan seaxes orde / hyd arypeð' ('Now some man wants to devour my flesh, gives no care for my pelt, after he rips the hide from my side with a knife's point,' ll. 4b-6a). Interestingly, the word used here is 'fretan' (though 'iteð' is also used later in the riddle). 'Etan' and 'fretan' overlap in their meanings, but 'etan' is used in broader senses, and 'fretan' has more meanings referring to animals eating things, and is also used to refer to consumption by fire. In German, these senses diverge completely, with 'essen' referring to human consumption of food, and 'fressen' used only for animals; the surviving record of Old English usage suggests traces of, or anticipation of, this distinction. The description in Riddle 77 of a person eating an animal using the word 'fretan,' alongside the narration of the human heedlessly ripping the 'fell' and 'hyd' ('pelt,' 'hide') from the oyster before devouring it, could have signaled to an Anglo-Saxon audience that the person was eating like an animal. The details recall Grendel ripping his victim limb from limb as he devours him, though Grendel's consumption avoids either 'etan' or fretan,' and is narrated instead using words that refer to the physicality of his actions, the embodied acts of consumption: 'bat' ('bit,' l. 742) and 'swealh' ('swallowed,' l. 743). The foregrounding of violence to the animal as a prelude to human consumption in Riddle 77, along with the monastic prohibition on eating mammals and the hermit Guthlac's renunciation of animal-based foods, suggests that the Anglo-Saxons had some sense that avoiding meat consumption was spiritually superior, though from the point of view of human asceticism rather than out of any concern for the animal.

Susan Crane argues that 'throughout their long history, the humanist traditions have tended to render nonhuman animals invisible to contemplation, unworthy of serious attention' (4). But the ways in which the Anglo-Saxons attended to animal natures, including the representation of animals in the *Exeter Book* riddles, whether the killing of a sheep for its skin or the appreciation of wild bird species for their distinctiveness, suggests that they may also have had a sense of animals as possessing a life force analogous to that of humans and thus warranting some esteem.

In narrating the death of the animals in their own imagined voices, and in describing the human killers as 'the enemy,' the riddles resist the cultural normalization of animal slaughter. Slaughter of a domesticated animal might not be an everyday occurrence, particularly among peasant farmers, though it would certainly have been common throughout the year. Hens and cows are more valuable for the milk and eggs they provide than for their meat, and sheep would be kept alive as long as possible for their wool, but the male offspring, like male lambs, would have been eaten. But the slaughter of domestic animals does not seem to be taken for granted in the riddles to quite the same extent as the consumption of meat from those same animals.

Animals, Humans, and Reason

Some of the *Exeter Book* riddles follow a prominent line of medieval thought in insisting that humans are divided from animals by the faculty of reason, and in establishing a priority of the reasoning human over the non-reasoning animal. The ecofeminist philosopher Val Plumwood describes a line of thinking beginning in classical texts and persisting with Descartes and Enlightenment thinkers forward through to deep ecologists – but unfortunately leaping over the Middle Ages with barely a reference – that insists on a dichotomy between reason and nature. Reason is associated with the masculine, with women being seen as less capable of reasoned thought and as tied more closely to nature through processes such as gestation and childbearing. Plumwood writes 'The concept of reason provides the unifying and defining contrast for the concept of nature.... The continual and cumulative overcoming of the domain of nature by reason engenders the western concept of progress and development' (1993: 3). Karl Steel has demonstrated that thinkers and creative writers in the later Middle Ages adopted this line of thinking from classical texts, and points out that it distinguishes not only between humans and animals but also between some humans and others. 'Medieval sources tend to separate these two groups [humans and animals] through an appeal to reason, a term that encompasses those capacities purportedly possessed only by humans, or at least those humans the text favors, whether philosophers, Christians, men, or noblemen' (2008: 11).

The same dichotomizing paradigm is articulated in Old English texts. According to the Old English Boethius, whose translation from Latin has been attributed to King Alfred, '...se mon ana hæfð gesceadwisnesse' ('the man alone has reason,' 1:317). In his 'Interrogationes Sigewulfi,' Ælfric comments in greater detail on what he perceives as the distinction between animals and humans:

> Hwi wolde god þæt se man Adam eallum nytenum naman gesceope? Þæt se man þurh þæt undergeate hu mycele betera he wæs þurh his gesceadwisnyssa þonne ða nytenu, & þæt he þæs þe swiþor his Scyppend lufode, þe hine swylcne geworhte.
> Why did God wish that that man Adam would shape names for all of the animals? So that, through that, the man should perceive how much better he was through his capacity for reason than the animals, and that he therefore would love the more his Creator, who created him in this manner. (13)

The *Exeter Book*, Riddle 47, typically solved as 'Bookworm' or 'Bookmoth,' works within this framework of insistence on the faculty of reason as that which establishes the boundary between animal and human and locates humans in a superior position:

> Moððe word fræt. Me þæt þuhte
> wrætlicu wyrd, þa ic þæt wundor gefrægn,
> þæt se wyrm forswealg wera gied sumes,
> þeof in þystro, þrymfæstne cwide
> ond þæs strangan staþol. Stælgiest ne wæs
> wihte þy gleawra, þe he þam wordum swealg.
>
> A moth fed on a word. It seemed to me a wondrous thing when I heard of the wonder, that the worm chewed on some song of man, majestic tale and strong foundation, a thief in shadow. The thievish spirit was not a bit the wiser, though he swallowed those words. (ll. 1-6)

The bookworm gains nothing from having swallowed words, because it does not understand human language. The worm does not have the human capacity of reason, thus differentiating it, an animal, from humans. Though it literally eats the words, it is unable to gain any knowledge or understanding from them.

While Riddle 47 ('Bookworm') asserts a clear boundary between the unthinking animal that feeds uncomprehendingly on the words of the book and the reasoning, spiritual person who would otherwise read it, Riddle 26 ('Book') blurs that boundary. Riddle 26 gives voice to the animal killed for its skin, against its will, to make a book, and complicates the easy assertion of a reason-based boundary between human and animals. The sheep, or perhaps cow, speaks in a voice that seems constant even as it narrates its own transitions through several states of being: from live animal into disembodied skin and then rectangles of parchment, written upon with the feather of a bird inscribing tracks of human language, and finally completed book decorated with gold and jewels: a sacred text that will bring good to those who use and love it. However, the utility of the book is not in the gold-adorned materiality of the leather volume that once served as skin for an animal, but in the inky tracks that have been left on the leaves, incomprehensible to the narrator because they are inscribed in the language of the human, whose voice ultimately silences that of the animal.

The riddle opens with nature in the form of an animal given voice, given agency, given the ability to complain about its own death:

> Mec feonda sum feore besnyþede,
> woruldstrenga binom, wætte siþþan,
> dyfde on wætre, dyde eft þonan,
> sette on sunnan, þær ic swiþe beleas
> herum þam þe ic hæfde.

> An enemy deprived me of life, robbed me of my strength, afterward wet me, dipped me in water and took me out again, set me in the sun where I quickly lost the hairs that I had. (ll. 1-5)

By the end of the riddle, the original animal narrator has become the book, and the substance of the book is not the physical materiality but the abstraction of the text it transmits. Thus, while it appears to give voice to nature, the riddle's anthropomorphizing narrative moves the animal from independent agent to subject of human dominance to material object, the physical materiality of which is irrelevant. What ecocritics have figured as a post-industrial narrative of the human domination of nature is inscribed in the surface of the skin that enables the dissemination of scripture within this riddle a thousand years old. Simultaneously, the riddle itself is inscribed in the skin that enables its transmission.

Riddle 26 simultaneously challenges the boundaries between animals, plants, and things. Animals and plants are used together by the human to create the book, which then becomes a thing, a dead thing, but read by a living human even after the death of the human creators of the book. Sarah Kay comments 'The act of writing comprises the touch of human skin on animal skin, goose feather pen in hand, oak gall ink in a horn inkwell close by; and reading involves renewing this contact of skin on skin, as the feather's traces are deciphered' (13). A 'gall' is a growth on a tree formed by insects living on it, so in addition to skin, feather, and horn, the book contains traces of insect and oak. As Holsinger points out, a manuscript is both 'a handwritten book produced by and for humans' and 'a stack of dead animal parts produced from and at the expense of animals' (619). The fact that the 'Book' riddle is voiced by an animal throughout forces attention to this.

Moreover, the animal refers to the human who kills it as 'enemy,' and uses the terms 'besnyþede' and 'binom,' both words capable of referring to robbery, an act of a human being against another human (or human institution): it is unusual outside of deep ecological contexts to think of it as possible to steal from nature. In conjunction with Riddle 77 ('Oyster'), which depicts the human as aggressive and even animal-like in its violent slaughter of the animal for food, Riddle 26 positions the animal as entitled to agency and consideration.

A few lines into the riddle, the 'excarnification of the animal' (Treharne 2013: 471) causes the living creature to give way to a stack of blank leaves of parchment:

Heard mec siþþan
snað seaxses ecg, sindrum begrunden;
fingras feoldan, ond mec fugles wyn
geond speddropum spyrede geneahhe,
ofer brunne brerd, beamtelge swealg,
streames dæle, stop eft on mec,
siþade sweartlast.

Afterward the hard edge of the knife cut me, sharpened with cinders; fingers folded me, and a bird's wing made abundant tracks of ink all over me, over the dark surface, swallowed the wood-dye, part of the stream, stepped often on me, wandered leaving black tracks. (ll. 5b-11a)

After the parchment has been prepared for writing, the feather – 'fugles wyn' ('birds' joy,' l. 7) dances over its surface and 'beamtelge swealg / streames dæle' ('drank in wood-dye, the stream's portion,' ll. 9b-10a). While the word 'swealg' can function metaphorically to mean that the feather absorbs the ink, the literal meaning is suggestive given that the parchment has so recently, within the narrative of the poem, been a living animal, and the feather makes sense as the kenning 'bird's joy' only if the bird is still alive. The word 'swealgan' is used to refer to literal eating in Riddle 47, in which the bookworm becomes none the wiser 'þe he þam wordum swealg' ('though he swallowed the words,' l. 6). The word 'sniðan' has an interesting range of meanings given this context: in addition to the general meaning of 'cut,' it can also denote 'amputate' and 'slay,' significations that recall the fact that the skin that is being written, and written on, came from an animal that was, only a few lines ago, alive.

The antlers of both Riddles 88 and 93 are used as inkwells. In Riddle 88, the antler is set upright on a board (likely placed in a hole in a writing surface). In Riddle 93, the narrator says 'Nu ic blace swelge / wuda ond wætre' ('now I swallow black wood and water,' ll. 23b-24a). Ink, as noted above, is made using tannic acid derived from oak gall, with ferrous oxide or soot added for pigment, hence 'black wood and water.' The use of the antlers as inkwells adds another body-part dimension to the creation of the Book of Riddle 26 (see Chapter 5): in addition to quills and skin, fingers, oak, and insect, the preparation of the manuscript requires a hollowed-out antler to hold the ink.

Many different people are responsible for the animal's transformation into a blank page, but the human involvement is reduced to an assemblage of disconnected body parts and tools: 'fingers folded me,' 'the knife's edge cut me,' 'a bird's wing [i.e. a quill] stepped on me': the hand holding knife and quill, and indeed the person to whom the fingers belong, are elided. The evocation of the different creatures involved in the production of the book – insect, tree, bird, sheep, deer, human – breaches boundaries among different living beings; the phenomenon of an animal given voice, and calling a human an 'enemy' likewise challenges the distinction between human and animal.

In Kay's reading of Riddle 26, 'This impossible transition of the speaking "I" from sheep to page captures the uncanny shortcircuit between animal, text, and book' (19). The Riddle further complicates the boundary between human and animal. As noted above, the slain animal says in line 1 that it loses 'feore.' When the sons of men are advised to read the words in the book, the riddle lists the advantages of doing so:

> Gif min bearn wera brucan willað,
> hy beoð þy gesundran ond þy sigefæstran,
> heortum þy hwætran ond þy hygebliþran,
> ferþe þy frodran, habbaþ freonda þy ma,
> swæsra ond gesibbra, soþra ond godra,
> tilra ond getreowra....
>
> If the sons of men will make good use of me, they will be healthier and more victorious, bolder in heart and more joyful in mood, wiser in spirit; they will have more friends, loved ones and kinsmen, true and good, brave and faithful. (ll. 18-23a)

As noted above, when the animal speaks in line 1, it says that an enemy has cut off its 'feore,' its life or spirit. This is echoed in line 21 when the narrator – still, somehow the animal/skin/book – says that those who heed the message written on it will be the wiser in their 'ferþe,' their 'soul, spirit, life' (Bosworth-Toller, *s.v.*). The creature-turned-book speaks in the language of the human about the benefits of heeding the human language inscribed on it; it recommends for the souls of – human – readers the text on behalf of which its own spirit has been cut short. Holsinger finds irony in the 'predictable disjunction between the comparative claims of dead animals and those of surviving books' (620). The book originates from the animal, but the text of the book originates elsewhere; as Elaine Treharne points out, 'there is more to 'text' than just words' (2013: 467). Riddle 47 ('Bookworm') makes that point, but definitively prioritizes the words over the materiality of the page. Riddle 26, conversely, problematizes that priority. In addition to words, the book requires flesh and feather, including the (living) flesh of humans that provide sustained and attentive labor of various kinds.

Riddle 26 is typically solved not simply as a book, but a book of liturgy or scripture. Within the context of Christianity, faith based on scripture yields the potential for eternal life, which transcends all earthly existence and negates any possibility of an animal without a soul being equated to a human being. This adds an additional layer to the riddle's interpretation, and this level of signification might be seen as the final word, moving beyond any other interpretation. Holsinger notes that 'the Bible that is the riddle's culminating solution, the "glorious" book that promises heaven, originates not from the words of the prophets, or from the inspiration of God, but from the flayed hide of the animal who gave its life and endured only as the ink-stained page of the book' (622). He further points out that 'mec' ('me') is the first word in the Riddle, and is repeated several times throughout, alongside 'me' ('me') and 'ic' ('I'). Riddle 26 continues to foreground the 'I' of the speaking animal, even as it shifts materially into something else. At the same time, auditors recognize that the animal whose skin is inscribed with the text of the book is incapable of understanding the scriptural message written on its own back. The sheep's reference to its killer as 'enemy' anthropomorphizes the animal and threatens its species distinction from the human, even as it reinforces the boundary between killer and killed. Yet the first line and a half of the riddle narrate the killing in the past tense: what actually speaks is not the live sheep, but the dead animal, apparently already skinned, as the next line describes soaking the skin as the first step in the preparation of parchment. Reading the riddle in terms of the object, as hairy pelt moving to prepared parchment to stitched and bound codex, makes it seem somewhat less unstable than conceptualizing movement from live animal to abstract textual entity. Yet as Bill Brown comments, 'however materially stable objects may seem, they are, let us say, different things in different scenes' (9). The animal's skin means something different on the back of the animal than it does as the leaves of the book, or wrapped around wooden panels used to bind the book and then decorated with twisted gold and jewels.

Val Plumwood challenges the classical dichotomizing of various categories, among them nature and human, arguing that instead of philosophical dualistic thinking, we need to understand nature and human as continuous, with shared features as well as differences. 'Overcoming the dualistic dynamic requires recognition of both continuity and difference; this means acknowledging the other as neither alien to and discontinuous from self nor assimilated to or an extension of self' (1993: 6). The 'bookworm' of Riddle 47 allows for dichotomous thinking, of sharp contrasts between human, animal, and object. The changing 'I' of Riddle 26, however, demands to be read in terms of continuities, contradictions, potentially incoherent or incomplete transitions. Within a dualistic framework, it is difficult to think about the shifting ontology of the riddle and to encompass the distance between its start- and endpoints. Riddle 26 enacts continuity between animal and book, between living being and inanimate thing, but it also narrates discontinuity between human and animal even as it gives human voice to the animal. The sheep is simultaneously voiced and silenced.

Riddle 14, 'Horn,' describes an animal reduced to one part of its body, in a treatment similar to the 'Book' riddle's descriptions of skin and feathers. Riddle 14 begins 'Ic wæs wæpenwiga. Nu mec wlonc þeceð / geong hagostealdmon golde ond sylfore, / woum wirbogum' (I was a weaponed warrior. Now a proud one covers me with gold and silver, twisted wire,' ll. 1-3a). Some editors and scholars would emend the word 'wæpenwiga' to 'wæpen wiga,' giving the reading 'I was the weapon of a warrior' (Muir 584). In this reading, the riddle's 'Ic' is the horn rather than the animal from which it is taken. Williamson takes this horn, and that of Riddle 80, to be 'great aurochs' horns similar to the drinking horns discovered at Sutton Hoo' (170), though previous scholars had proposed falcon, sword, or spear (Muir 688, Tupper 217).

Riddle 26 ('Book') is analogous in beginning with the animal which, during the course of the riddle's narrative, becomes parchment and then book, and gives context for reading Riddle 14 as beginning as an animal from whose head a horn has been removed, and then moving to the horn, and then to the horn's use as a war-horn or drinking vessel, carried on a horse or a ship. When not in use, the horn might hang as treasure on a wall; or 'hwilum ic bordum sceal, / heard, heafodleas, behlyþed licgan' ('sometimes I must lie on a hard board, headless, deprived,' ll. 9b-10). The horn is 'headless' because it has been removed from the head of an animal, though the description is unexpected; an audience might expect a body to be 'headless,' not something that came off the head itself. The reversal serves the trickery of the riddle while simultaneously asserting a continuity between horn and body. Riddle 80, also solved as 'horn,' has a harsh voice ('heard is min tunge,' l. 8b); it rides on a horse, and it holds within it the product of the grove, i.e. mead made from honey. As Holsinger notes of Riddle 26 ('Book'), repetition of 'ic' and 'mec' throughout both Horn riddles 'holds up an ethical mirror to the centuries of slaughter' (622) that leave the richly decorated horns that still survive in museums.

Much as the sheep is reduced, yet not reduced, to its skin, the two Horn riddles recall the animal from which they were taken while simultaneously describing the horns as objects, independent items rather than parts of animals now dead. Aurochs remains can be found in Great Britain in Neolithic sites, but by the Anglo-Saxon period these animals had long been extinct. Rare examples of aurochs horns from the Anglo-Saxon period do exist, as for example from the Sutton Hoo excavation, but these must have been imported from the Continent, where aurochs lived on for another several centuries, dying out last in eastern Europe (Banham and Faith 85). The horns, then, are doubly displaced, recalling their existence as living animals yet identifiable within the context of Anglo-Saxon England only as objects brought from afar. The fact that aurochs are wild animals also differentiates them from the domesticated sheep of the Book riddle.

Animal Aesthetics and Agency

Several of the *Exeter Book* riddles describe birds, and one a fish, in terms very different from those found in *Beowulf* and other heroic poems, which present animals and humans in a starkly hierarchical framework, with formulaic language for animals alongside direct statements of human priority over animals. As discussed in greater detail in Chapter 2, Beowulf slaughters animals he refers to variously as 'hronfixas' ('whales,' l. 540), 'merefixa' ('sea-fish,' l. 549), 'fah feondscaða' ('hostile enemy,' l. 554), 'aglæcan' ('monster, fighter,' l. 556), 'mihtig meredeor' ('mighty sea-animal,' l. 558), and 'niceras' ('sea-monsters,' l. 575). A few lines later, Beowulf uses the term 'æglæca' (l. 592) to refer to Grendel. Nearly three dozen words for different species of fish are attested in the surviving Old English corpus. In these lines, Beowulf is not aiming for a description of a specific fish; he describes sea-monsters in language used interchangeably for the land-dwelling, humanoid monster Grendel. The monstrous sea-creatures function to show off Beowulf's prowess simultaneously in his past contest with Breca and in his battle of word and wits with Unferth. Although the fish belong in and to the sea, and Beowulf can be said to have invaded their home with Breca and his naked sword, he justifies their killing by claiming: 'syðþan na / ymb brontne ford brimliðende / lade ne letton' ('after this, about the steep ford, [they] would not hinder seafarers' journeys,' ll. 567b-69). This is in marked contrast to the description of fish and river, or of wild birds carefully observed in their own habitats, in the *Exeter Book* riddles

From an ecocritical point of view, the 'beasts of battle' found in nine Old English poems bear a resemblance to the 'sea-beasts' that Beowulf encounters during the swimming contest, in that the animals are described in formulaic language and are subordinated to human concerns. Like the 'sea-beasts' slain by Beowulf, the 'beasts of battle' are undifferentiated and functionally equivalent to one another. Moreover, they appear within the narratives not out of intrinsic interest in their presence but in structural and/or metaphorical relationships to human actors within the poems. As the human warriors gather and prepare to fight, the beasts of battle circle, awaiting human carrion on which to feast. In *Beowulf*, a messenger tells of Beowulf's death, and predicts that neighboring tribes will now attack, and the beasts of battle will come to feast on the dead:

> se wonna hrefn
> fus ofer fægum fela reordian,
> earne secgan hu him æt æte speow,
> þenden he wið wulf wæl reafode
>
> The dark raven, greedy for those fated [to death] will make abundant noise, tell the eagle how he fared at eating, while, with the wolf, he plunders dead bodies. (ll. 3024b-27)

Eight other poems contain very similar 'beasts of battle' passages. *Genesis A*, one of two verse adaptations of portions of the biblical Genesis, adds a 'beasts of battle' passage to the description of the battle between the four kings of Sodom and Gomorrah, with no analogue in the biblical text: 'Sang se wanna fugel / under deoreðsceaftum, deawigfeðera, / hræs on wenan' ('The dark bird, dewy-feathered, sang under the javelin shafts, expecting carrion,' *Genesis A*, ll. 1983b-85a). A similar passage is added to the Old English verse adaptation of *Exodus*, at the point at which Pharaoh's army is preparing to attack the fleeing Israelites:

> Hreopon herefugolas, hilde grædige,
> deawigfeðere ofer drihtneum,
> wonn wælceasega. Wulfas sungon
> atol æfenleoð ætes on wenan
>
> Birds of prey screeched, greedy for combat, dewy-feathered over the bodies of the slain, dark carrion-pickers. Wolves howled an awful even-song in expectation of food. (ll. 162-65)

The phrase 'beasts of battle' does not appear in Old English poetry. In a study of the form and function of 'beasts of battle' passages in Old English poems, Thomas Honegger notes that the phrase was first used in 1955 by Francis Magoun, who called it 'an ornamental rather than an essential theme' (Honegger 83). The idea that animals are mere decoration, with no essential meaning for humans or function in the poetry, is contradicted by their obvious importance. In the formulaic collocation, these animals lack differentiation; they are treated as a group of non-human creatures whose presence in the poem has to do not with their independent existence as eagle or hawk or wolf but with their utility in metaphorical terms to define the human, in this case the human male as warrior preparing for battle and, presumably, slaughter of other humans.

The deployment of the 'beasts of battle' in these poems is complicated by the multiplicities of meaning attributed to them. In that they are predators who seek live prey to kill it and eat it, they stand in metaphorically for the humans readying themselves to kill. The 'beasts of battle' are explicitly described as circling in anticipation of food, however, and in this sense they are not hunters, but scavengers in search of already killed meat. Eagles, hawks, and wolves do in fact scavenge for carrion, in addition to preying on dead animals, but such activity does not seem to befit the savagery that is their metaphorical force here. As scavengers who feed on the dead humans, the beasts of battle also pose a problem: the humans killing each other will 'scavenge' among the corpses of the dead for weapons, armor, jewelry, and other valuables, which then becomes part of the economic system of the society in question, as in Felix's *Vita Guthlaci*, though Guthlac is instructed to return a portion of the plunder in an acknowledgement that the practice is problematic (see Chapter 4). But where humans rely on other humans for food, as in the case of the Mermedonians in *Andreas* or, possibly, of Grendel, if he can be considered human, they become cannibals, monsters. The slipperiness in the 'beasts of battle' metaphors between slaughter and scavenging contributes to a slippage in the line between human and animal.

Many of the poems that include 'beasts of battle' passages present them in implicit comparison to human warriors. *Elene* makes this comparison more direct, by intertwining references to the battle-song of the wolf and the eagle, the mustering of human warriors, and the circling raven eager for carrion, highlighting the literal and metaphorical equivalences between humans and wild animals. The passage is worth quoting at some length:

For folca gedryht.	Fyrdleoð agol
wulf on wealde,	wælrune ne mað.
Urigfeðera earn	sang ahof,
laðum on laste.	Lungre scynde
ofer burg enta	beaduþreata mæst,
hergum to hilde,	swylce Huna cyning
ymbsittendra	awer meahte
abannan to beadwe	burgwigendra.
For fyrda mæst....	

Wæron Romware,
secgas sigerofe, sona gegearwod
wæpnum to wigge, þeah hie werod læsse
hæfdon to hilde þonne Huna cining;
ridon ymb rofne, þonne rand dynede,
campwudu clynede, cyning þreate for,
herge to hilde. Hrefen uppe gol,
wan ond wælfel. Werod wæs on tyhte.

The army of the people moved forward. The wolf in the wood chanted a war-song, did not hide the coming slaughter. The dewy-feathered eagle lifted up a song on the path of the hated ones. The largest army of enemy warriors that the king of the surrounding Huns could possibly assemble quickly drove forward in battle against the ancient city. The largest army set forth.... The Romans, victory-renowned men, quickly readied their weapons to battle, although they had fewer troops to go to war than the king of the Huns; they rode around the renowned one, then the shield rang and shield clanged; the king pressed forth with his army to battle. High above the raven cried, dark and slaughter-greedy. The army was in motion. (ll. 27-35a, 46b-53)

The description of the mustering of troops is framed by references to the raven, the eagle and the wolf circling in anticipation of carrion, with both armies related metaphorically to animals. Like other 'beasts of battle' passages, this one describes the animals as greedy for carrion, suggesting that they are present to feast on the bodies of fallen human warriors after the battle. But the juxtaposition in this passage between carrion-greedy beasts and war-ready humans suggests also a symbolic connection between the animals and the humans. The ferocity of eagle, hawk and wolf as predators in search of live food, rather than as scavengers feeding on already slain carcasses, is set against the eagerness of human warriors for battle. As careful observers of wild animals, as can be seen from the *Exeter Book* riddles, Anglo-Saxon audiences would have been aware that predatory animals also scavenge for carrion, though modern city-dwellers might think of them in sharp distinction to birds such as vultures that eat only carrion and do not hunt live animals.

The metaphorical play suggested by the beasts of battle, then, is complex. The animals themselves are presented as interchangeable, the two birds and the mammal all providing figurative fodder for author or scribe in enriching the characteristics of the human warriors, suggesting through association with these animals the humans' wildness and ferocity. Menely and Ronda argue: 'Rather than ritualize violence, modern societies place it at a distance' (27). They suggest in passing that pre-modern societies create rituals of violence either through animal sacrifice or through the elaborate rules that structure the hunt such as emerge in the later Middle Ages. There are, however, no surviving Anglo-Saxon texts that depict either animal sacrifice or the ritualized hunting of later medieval texts. War is not placed at a distance. The battles of *Genesis* and *Exodus*, while distanced in place and time from Anglo-Saxon England, are linked through language and imagery to the battles described in poems about local English conflicts such as *The Battle of Maldon* (Estes 2007). Battle-gore is often described quite explicitly. The 'beasts of battle' scenes provide some structure for the violence of battle, described in ways at once realistic and ritualistic.

In his objection to the treatment of animals as undifferentiated, Derrida might point to the use of the 'beasts of battle' as an instance of the assumption that animals are singular in their distinction from humans. He insists that readers acknowledge that there is no single totality that can be designated 'the animal,' but rather 'a heterogeneous multiplicity ... of organizations of relations [that] do not leave room for any simple exteriority of one term with respect to another' (2002: 399). In contrast to the hierarchical depiction of animal-human relationships in epic and religious poems, the *Exeter Book* riddles depict birds in ways that suggest rhizomatic connections among them, more lateral than hierarchical.

Several of the riddles describe wild birds using almost completely distinctive language and descriptive details reflecting the varied qualities of the birds and the close observation humans have made of them. Riddle 7, 'Swan,' distinguishes between the silence of the bird when walking on earth or floating on water, and the sound the bird's wings make when flapping:

> Hrægl min swigað, þonne ic hrusan trede,
> oþþe þa wic buge, oþþe wado drefe.
> ... Frætwe mine
> swogað hlude ond swinsiað,
> torhte singað, þonne ic getenge ne beom
> flode ond foldan, ferende gæst.

> My clothes are silent when I walk on the earth, or sink into my home, or swim in the water ... My adornment rustles loudly and sounds, sings clearly, when I am not at rest, on water and earth, flying spirit. (ll. 1-2, 6b-9)

This riddle comments on the sounds made by the swan's wings in flight. In contrast, Riddle 8, 'Jay,' Riddle 24, 'Magpie,' and Riddle 57, perhaps a 'flock of sparrows,' also comment on the sounds made by the birds, but in each case to the birds' song rather than the sound of their flapping wings; each of the riddles used distinct language to comment on this. The bird of Riddle 8 begins 'Ic þurh muþ sprece mongum reordum, / wrencum singe, wrixle geneahhe / heafodwoþe, hlude cirme' ('I speak through my mouth with many voices, sing with deceit, switch abundantly my voice, loud cry,' ll. 1-3). Riddle 24 also speaks in the voice of the creature whose identity is to be guessed, but rather than simply stating that it constantly changes its tune, it lists several of the other animals that it is known to imitate:

hwilum beorce swa hund, hwilum blæte swa gat,
hwilum græde swa gos, hwilum gielle swa hafoc,
hwilum ic onhyrge þone haswan earn,
guðfugles hleoþor, hwilum glidan reorde
muþe gemæne, hwilum mæwes song.

Sometimes I bark like a dog, sometimes bleat like a goat, sometimes cry out like a goose, sometimes yell like a hawk, sometimes I imitate the dusky eagle, battle-bird's cry, sometimes the vulture's call, with a wicked mouth, sometimes the song of the sea-gull. (ll. 2-6)

The language here is again completely different from the language of Riddle 8. Riddle 57, solved as a flock of birds of some kind, contains a description of the birds' sound completely different from that of the previous riddles, describing their song as something of a constant chatter: 'Sanges rope / heapum ferað, hlude cirmað' ('Liberal with song, they travel in flocks, shrieking loudly,' ll. 3b-4).

Two further Riddles, 9 and 10, have solutions having to do with birds, but make no reference to the sound or song or silence of their subjects. Like the other bird riddles, these are narrated in the voices of the birds themselves, and describe specific features particular to each individual bird. The cuckoo lays its eggs in the nests of other birds, which then incubate them and feed the resulting chicks, as described in Riddle 9: 'Mec seo friþe mæg fedde siþþan, / oþþæt ic aweox, widdor meahte / siþas asettan' ('The stately woman fed me afterwards, until I grew, and could set off on wider journeys,' ll. 9-11a). Finally, Riddle 10, commonly solved as 'Barnacle Goose,' describes a bird that migrates to England as an adult, so that its eggs and chicks were never seen, giving rise to legends that these birds grew to maturity in the water:

> Neb wæs min on nearwe, ond ic neoþan wætre,
> flode underflowen, firgenstreamum
> swiþe besuncen, ond on sunde awox
> ufan yþum þeaht
>
> My beak was confined, and I under water, overflowed by flood, sunk under mountain streams, and grew up in the sound, concealed from above by the waves. (ll. 1-4a)

Like the other 'bird' riddles, this passage describes its subject in detail that identifies the creature not as some a generic animal or beast, but as a specific species of bird.

Notably, these riddles include no reference to humans. There is no discussion of contact with humans, no use of birds in metaphorical terms that liken them to humans, no anthropomorphizing likeness between birds and humans. They are simply birds in nature, somewhat mysterious, creatures of interest in their own terms. In *Beowulf*, as argued above, animals are killed as a demonstration of the hero's physical, perhaps superhuman, strength and his martial prowess. In the *Vita Guthlaci*, the prose *Guthlac*, and *Guthlac A*, animals demonstrate Guthlac's devotion, his dominion over them as both symptom and proof of his spiritual elevation over landscape and animals as well as Britons and other humans. In these riddles, however, the birds serve no instrumental connection to humans. The riddles are told as entertainment; the birds are described in tricky detail so as to be identifiable, but not too easily so.

Importantly, the birds of these riddles are all wild animals, not domesticated ones; they pose no danger to humans or livestock, though they might need to be kept away from recently sown fields; they are of limited potential use to humans for food or other purposes. The close descriptions of them reflect careful observation and detailed knowledge in the complete absence of any utility to humans. Ecocritics who assume that appreciation for the landscape is post-medieval, or requires an urban sensibility, as discussed in Chapter 3 on ruins, might be surprised that the 'other' people who occupy those pastoral landscapes can also perceive them, can appreciate them for their own sake and on their own terms. They might be even more surprised to discover that even in the 'dark' ages, people could perceive nature around them, even when it was not of use to them, and describe its beauty and its particularity.

Perhaps even more remarkable is Riddle 85, in which a fish describes its life in a river and tells the audience, 'unc dryhten scop' ('God created the two of us,' l. 2). The fish adds: 'Ic him in wunige a þenden ic lifge; / gif wit unc gedælað, me bið deað witod' ('I always dwell in it, as long as I live; if we separate, death is certain for me,' ll. 6-7). The fact that Old English had grammatical as well as natural gender complicates the line, because the fish literally says, 'I live in him,' using a pronoun that can refer to an object or in a human being. (The Old English words for 'river' include 'ea' and 'lacu,' both feminine in gender, as well as 'flod,' 'stream' and several '-stream' compounds, which are masculine.) Writing about the bird riddles and *The Phoenix*, Donna Beth Ellard argues that they 'characterize ecosystemic partnerships between Anglo-Saxons and domestic and wild birds not as thought-acts but as intentional and creative encounters that make it difficult to distinguish between subject and object or among gamer, riddler, and bird' (Ellard 274). Likewise, in the fish/river riddle, the fish, an animate creature, does not give itself priority over the non-living river, but presents itself in equivalence with the river. The word 'gedælað' has the general meaning of separating, dividing, or distributing separate objects, but also refers to the division of a whole into parts. The fish and the river are distinct from one another, but are simultaneously two parts of a whole. If separated from the river, the fish will die; if separated from the earth, humans cannot live. The riddle imparts a sense of wholeness in creation, a sense that all of the world, the planet and the plants and animals and humans that live in and on it, are a single entity as God's creation. It simultaneously suggests a non-theological lived reality that anticipates modern environmental understanding of the interconnections among animals, humans, and landscapes.

Ellard argues that 'Isidore's avian etymology resonates as Deleuzoguattarian refrains that take birds across and beyond places that can be accessed by humans or identified by human semiotics' (275). Deleuze and Guattari argued that we need to develop non-hierarchical relationships among people and others, suggesting instead as a metaphor the rhizome, a root system that grows and branches continuously, and not insignificantly, below the ground (5). The *Exeter Book* riddles suggest rhizomatic or Quinean web-like relationships among their subjects and objects, rather than ones characterized by ranking, stratification, or dichotomies, not least because birds and fish can go places that humans cannot go: we can neither fly nor live under water. Birds and fish provide a reminder that much of the planet is uninhabitable for humans. This is a source of anxiety for Beowulf, who swims through the sea slaughtering the animals he encounters, later protesting that he has made the seas safe for travelers. But for the scribes and audience of the *Exeter Book* riddles, the voices of fish and birds provide a reminder that humans are part of a complex world with multiple perspectives, potentially in harmony, though also potentially in competition, as when humans are threatened by storms or animals, or when animals refer to humans as 'the enemy.'

In the riddles, animals are depicted in detail specific to their species; these include bivalves, birds, and mammals, wild as well as domesticated animals, dwellers on land and sea. There is no such thing in the riddles as 'an animal,' and certainly no 'the animal'; particular species of animal are described using specific details that simultaneously confound the listener and subtly reveal the answer. In contrast, the 'beasts of battle' passages obscure differences among eagle, raven, hawk, and wolf, using interchangeable language to describe their sounds and their greed for carrion, subordinating them in symbolic language to human concerns.

Two riddles describe objects made from antlers, figuring them as having lost their homeland, which is, broadly speaking, in the forest, but also on the heads of deer, from which they have been disenfranchised by cyclical, annual cycles of loss and regrowth. The transformation of antler into inkwell or drinking cup is interestingly distinct from the use of skin to make a book or horn to make a drinking vessel or a musical instrument, because the animal does not need to be killed to acquire the raw material: deer shed their antlers each year in late winter or early spring and then grow new, larger ones. While casual hikers are unlikely to stumble over them, a dedicated search in areas where animals are known to feed or sleep is reasonably likely to yield the discovery of a rack of shed antlers. Unlike sheep, deer are wild animals; unlike the beasts of battle, deer are not predators; they eat grass and leaves. Yet the males fight among themselves to establish territorial dominance or for the attentions of a female, and the large antlers of older animals suggest masculinity, prowess, martial success – hence Hrothgar's name for his high hall, 'Heorot,' i.e. 'hart' or 'stag.' Antlers can be taken from a deer that has been hunted and killed. However, the 'Antler' riddles do not refer to killing deer to take their antlers; rather, they recall their time living on the heads of live deer. Riddle 93 refers to the new growth that replaces it after it falls off the stag's head after the annual shedding of antlers as 'younger brother': 'Nu unc mæran twam magas uncre / sculon æfter cuman, eard oðþringan / gingran broþor' ('now two more kinsmen of ours shall come after us, deprive us of our dwelling-place, younger brothers,' ll. 14-16b). Riddle 93 thus spells out clearly that the antler has not been taken from an animal that has been killed, but has fallen in the normal course of the year from the animal's head to be replaced by another.

The opening lines of Riddle 88 are badly damaged. The first line begins, 'Ic weox þær ic s....' ('I grew where I ...,' l. 1), and of the subsequent lines only a few words survive, so it is impossible to know if the speaker begins as the animal, as with Riddles 26 ('Book') and 14 ('Horn'), or if it begins as one of a pair of antlers already separated from the animal, as it describes itself in line 12 and following. As antler, the speaker describes itself along with 'broþor min' ('my brother'), as protected by the forest: 'Ful oft unc holt wrugon, / wudubeama helm wonnum nihtum, / scildon wið scurum' ('Very often the woods shielded us, beams of wood shielded us against showers, our helmet in the dark night,' ll. 11b-13a). As the riddle-narrative progresses, the antler soon becomes a solitary being: 'Nu... eom ic gumcynnes / anga ofer eorþan' ('Now, I am alone on earth among men,' ll. 14a, 16b-17a). The Antler repeats the idea of having been twinned with a 'broþor' and also uses the dual pronouns 'wit' and 'unc' several times. More than its utility as a human object, the riddle focuses on the antler's lost relationship with its twin, evoking relationships among humans and making porous the border between human and animal, and between animal and object.

Riddle 93 also describes one antler from a pair that once lived on the head of a stag, but rather than focusing on the 'brother,' the riddle describes the life of the animal in the forest. As in Riddle 88, the speaker of Riddle 93 states explicitly that it has fallen off the head of the stag and been replaced by new growth: 'Ic on fusum rad / oþþæt him þone gleawstol gingra broþor / min agnade ond mec of earde adraf' ('I rode eagerly, until my younger brother possessed the seat of joy and drove me from my home,' ll. 13b-15). 'Eard' ('home,' l. 15) also has meanings of native country, land, province (Bosworth-Toller, *s.v.*), suggesting the antler's unwilling dispossession and movement to a place it has not chosen, displaced by another being. In suggesting that the antlers are alienated from a lost home, these riddles suggest that people might nostalgically recall a lost past, such as is evoked more directly in poems such as *Beowulf* and *The Ruin*, discussed in Chapter 4. The idea that the deer (or its antlers) belong to a particular 'earth' is an interesting concept for a people that defined itself around the idea of migration, registering the idea that a given population might have a legitimate claim to a particular territory simultaneously with the notion that, as animals, the claim of the deer would be less legitimate than the use made by humans, in an erasure of legitimacy analogous to that made through the slippage between Britons and demons in the narratives of Guthlac (see Chapter 4). The chapter that follows details other things taken from their 'home' in the earth, including trees and ore.

Conclusion

Old English theological texts insist on a clear distinction between human and animal; the depiction of the sea-beasts in *Beowulf* and the variety of small wild animals in the Guthlac narratives also points to the idea of a firm line between humans and animals. The metaphorical use of animals in the 'beasts of battle' type-scenes depends upon an understanding of human and animal as fundamentally distinct, because otherwise they open uncomfortable possibilities for interpretation. Yet at the same time the intertwined description of 'beasts of battle' and human warriors preparing for battle in *Elene* suggests an awareness of analogues and similarities between humans and animals. In the *Exeter Book* riddles, the depiction of the fish that cannot live except in the environment of the river suggests awareness of human dependence upon the natural environment. The observant, detailed descriptions of wild birds in the riddles, used in playful appreciation, notably contrast with the appropriation of animals in *Beowulf, Elene, Exodus*, and other heroic poems to articulate human desires and needs, whether metaphorically or directly.

Scholars of environmental issues have noted the problems that arise from privileging human perspectives in literary and other texts. Even some environmental writing has seen animals as worthy of attention only insofar as they can be understood to resemble humans, or their extinction as an index of human loss. Karl Steel argues that animal liberation efforts have foregrounded human concerns so that they 'undermine their own goals by measuring animals against idealized human capabilities to determine which creatures merit consideration as ethical subjects' (Steel 2011: 3). Object-oriented ontology and its materially-focused cousins, as well as critical animal theory, instead see humans as interconnected with animals, with plants, and with objects in ways that allow for, even insist upon, an ethics of difference, as discussed in more detail in the following chapter. The *Exeter Book* riddles do important work in challenging anthropomorphizing readings about environments, about the plants and animals and objects, man-made and otherwise, that constitute those environments. In discussing wild birds with careful attention to details about plumage, habits, and sounds, but without reference to human concerns, these riddles suggest that such birds are worthy of attention simply because they exist. In giving voice to bird and fish, to sheep and book, the riddles open the possibility for a recognition that human perspectives are limited and limiting.

Riddle 26 ('Book') speaks for the dead animal in the voice of the human. It subsumes the dead animal's skin into a codex, an object to be venerated for its textual contents, yet the opening cry, 'Mec feonda sum feore besnyþede' ('an enemy robbed me of my life') projects the human as enemy and resonates through to the end of the riddle, forcing a recognition of the agency of the animal even as it has been silenced under the metaphorical weight/

freight of the scripture with which it has been inscribed. Jeffrey Jerome Cohen argues that though humans possess the capacity for reason, we often nevertheless act unreasonably. 'What is at stake in limiting agency to an origin in human volition – as if we intend much of what we accomplish? The profundity of climate change in the anthropocene argues against such easy alignment' (2013a: xxiv).

Derrida notes that the notion of a boundary between human and animal, and the ways in which humans have imagined that boundary, 'has a history' (399). This chapter attempts to limn a moment in that history, a moment with traces left behind in a series of poems written in an ancestor of today's English language, with words the connotations and denotations of which are at least partially lost to us. Ecological thinking also has a history, and as Susan Fraiman has noted, ecofeminist activist-philosophers from the 1970s have sometimes been ignored in that history in favor of 'posthumanist' philosophers who trace their roots to Derrida instead. 'Those mobilizing Derrida typically distinguish their project not only from animal advocacy but also from gender studies and other areas animated by specific political commitments' (90). A commitment to advocacy lies behind this book, however, and in the chapters that precede and follow this one, I make the effort to connect environmental issues with political ones of colonization and gender.

While the riddles enact the subjugation of animals through violence and through the claim of lack of reason, they also provide an alternative vision whereby animals protest the violence that is done to them, using human language to protest and to designate humans as 'the enemy.' Yet this occurs in the context of the riddles, designed to mislead the hearer into thinking about something other than the solution. It could be quite cogently argued that any riddle that appears to give reason or thought to an animal does so precisely because animals do not have these qualities and therefore they underline this lack rather than arguing against it. Yet as Freud argued in *Jokes and their Relation to the Unconscious*, there is serious meaning to all foolery, including that of the riddles. The riddles were composed and written down by humans, using the skins of mammals, the feathers of birds, tannins derived from insects nesting in the bark of trees, and pigment from the ashes of wood or from rusted ore: the possible message that there are limitations in the possibilities of human voice are transmitted by humans, using materials derived from the deaths of animals, and to indulge human wit. The message is fraught with conflict.

The following chapter continues to engage with the *Exeter Book* riddles, discussing once again Riddle 26 ('Book') because it moves from living animal to object to leaves in a book. Numerous other riddles concern things, some made from living plants, some from ore drawn from the earth, but speak in voices that protest human treatment. Other riddles place humans in the position of objects. The chapter draws on philosophical meditations on the status of things: thing theory, object-oriented ontology, and hyperobjects. Things are parts of landscapes, are excavated or cut from landscapes, and form an essential part of the ecological world, though they do not seem 'environmental' in the same way as do sublime mountain peaks, beautiful flowers, or malodorous marshlands.

6 Objects and Hyperobjects

Introduction

The previous chapters have considered human relationships with other humans, individually as well as in groups, as well as with animals and landscapes, the more 'traditional' topics of environmental criticism. This chapter turns toward things, using recent philosophical insights in the areas of thing theory, object-oriented ontology, and hyperobjects to consider how objects interact with living beings, and vice versa, in the *Exeter Book* riddles and in the culture in which they were written. Thinking about the many objects described in the riddles, and the natural materials of which they are made, complicates what we think we mean by the word 'object' and how 'objects' relate to, and are related to, humans. The riddles do not simply *describe* things, animals, and people, but they transmit ideas and ideologies, cultural assumptions as well as potential resistance to such assumptions, about the roles and functions of animals and other living things, as well as non-living parts of the non-human world. The objects described in many of the riddles provide an opportunity to think as if humans are not the center of consideration, but pushed to the edge, with things made central.

In animal studies, the argument has been made that animals deserve moral consideration because, like humans, they suffer. One problem with this ethical orientation is that it maintains the human at the center: animals are seen as worthy of moral consideration and individual agency insofar as they resemble humans. Another is that it fails to provide a rationale for the ethical consideration for beings that do not have nervous systems or feel pain in ways analogous to humans. Can the landscape feel pain? Receding glaciers, flooding and drought, forest fires and rising sea levels suggest that perhaps the earth is suffering, but it seems rather problematic to suggest that it suffers in the same sense as a sentient being. Yet ecofeminism, object-oriented ontology, thing theory and related philosophical endeavors make the claim that inanimate objects also warrant ethical consideration parallel to that of humans, not because of similarity but simply because they exist. As Jeffrey Jerome Cohen comments in the introduction to *Animal, Vegetable, Mineral*, 'the study of animals, plants, stones, tracks, stools, and other objects can lead us to important new insights about the past and present; ...they possess integrity, power, independence and vibrancy.' (2014: 7).

The *Exeter Book* riddles anticipate recent theoretical explorations by giving voice to objects. Like Riddle 26 ('Book') discussed in the previous chapter, these riddles speak in shifting voices, moving from starting points such as ore deep in the earth or trees under the sky, through the processes of mining or killing them, to descriptions of objects made from them, with no clear or fixed point of rest. The riddles point to multiplicity rather than allowing for singularity or stability. Philosophical examinations of the moral agency of things connect with ecocritical perspectives that call for attention to landscapes and other non-sentient beings, not for the sake of human occupants but simply because they exist. While some of the riddles about objects maintain the centrality of the human, others push humans to the margins, where they occupy a status secondary to that of the objects that describe themselves. Even given that human subjects and objects are quickly restored by language or riddling context to the focus of consideration, the possibility of centering an object anticipates in fascinating ways not only the work of object-oriented philosophers but also the argument of Patricia Hill Collins that the center must always be in motion, allowing for constantly pivoting points of reference.

Thing theory distinguishes 'objects' – items worthy of display in museums – from 'things' – everyday items. Object-oriented ontology (hereafter OOO) places humans on par with all other things, ascribing potentially equal importance to elephants and icebergs, staples and the Queen of England. Timothy Morton invented the term 'hyperobject' to refer to things that are too large to perceive in their entirety, like all the Styrofoam in the entire world, or climate change. The idea of the hyperobject is also useful in thinking about social structures such as racism and sexism, and intersects interestingly with some of the things (not 'objects') described in riddles, as explored in the second half of this chapter. Poems such as *Beowulf* or *Andreas*, alongside prose works such as King Alfred's 'Preface' to the translation of Augustine's *Soliloquies*, take animals and rocks and trees for granted as materials for human consumption. The riddles include things and objects as varied as icebergs and storms, weapons and chalices, items of utility or of spiritual symbolism that are made from materials that describe being torn from their 'homelands' by human enemies. Wood and stone voice subjectivity and thus make a claim for ethical consideration and agency.

Decentering the Human

Many of the *Exeter Book* riddles are written from the point of view of a non-human speaker, with an object or animal describing itself beginning with 'ic' ('I') and concluding with some variation of 'say what I am called.' In addition, there are several that describe themselves as having been torn away by human enemies from some kind of homeland. In Riddle 93 ('Antler'), examined in the previous chapter, the antler speaks of having had its place on 'eard' ('earth,' 'homeland') stolen by a younger brother – the new antler that grows in after the stag sheds the previous year's growth.

In seeking an ethical reading of things, Bill Brown asks 'What are the conditions ... for sympathizing with animals and artifacts, and how does such sympathy threaten Locke's "thinking thing," the self?' (7). Locke's self, of course, post-dates the *Exeter Book* riddles by several centuries, but the Anglo-Saxons, too, had a sense of self; they imagined the chest cavity as the center of emotion and reason. The *Exeter Book* riddles imagine sympathy for things in conjunction with a human sense of self. Plumwood suggested an answer to the question of how sympathy for objects challenges human subjectivity in her argument challenging dualisms such as those posed between human and animal, thing and object, and indeed challenging the very category of 'other'. Plumwood rejected dualistic thinking altogether, in a series of publications with profound implications for ecocriticism but unfortunately given too little attention in more recent ecocritical studies. She points out that 'nature must be seen as a political rather than a descriptive category' (3), and thus the distinction between human and nature (or, as ecologists often call it, 'non-human nature') is a distinction constructed and renewed by social norms and practices rather than one that exists 'in nature.'

In Riddle 35, usually solved as 'Ore,' the speaking subject says that it was born from the earth: 'Mec se wæta wong, wundrum freorig, / of his innaþe ærist cende' ('The damp earth, wondrous cold, first bore me out of his innards,' ll. 1-2). The idea that stone can be birthed suggests a parallel with humans and animals. Adam, the first man, is also said to have been born from earth, in another riddle found only in MS. Cotton Tiberius A. iii among a list of biblical curiosities and lore:

> Hwæt wæs se on þissere worulde seðe acænned næs, 7 þeah hwæðere wæs to men geworden 7 lange lifde? 7 þa eft æfter his deaðe þæt he wæs bebyrged innon his modor innoðe? 7 æfter þam deaðe eft þæt hit gelamp æfter manegum wintrum þæt he wæs gefullwad, 7 næfre his lichama ne fulode ne ne brosnode innon þære eorðan? ... Þæt wæs Adam, se æresta mann þe þis bigelumpen wæs.
>
> Who was he, who in this world was not born, and nevertheless was made among men and lived long? And then again after his death, he was buried

> inside his mother's womb? And again after death, it occurred after many winters that he was baptised, and his body never fouled or decayed in the earth? ... That was Adam, the first man, that this happened to. (Estes 2012: 643-44)

Adam's 'mother's womb' is the earth, from which he was made. The source of the ore is, likewise, the earth, so that the 'brother of earth' can be understood as the brother of ore: human and stone are siblings.

The Cotton Tiberius riddle uses 'acænnan' and 'innoð' with the feminine noun 'eorðe,' while Riddle 35 uses 'cennan' and 'innaþ' with the masculine noun 'wong.' 'Cennan' and 'acænnan' have very similar ranges of meaning. Both refer to bearing a child, being born, begetting a child, as well as to raising plants, and both also have a meaning connected with proclaiming or making something known through speech. Both words can refer to the mother's role in giving birth as well as to the father's role in conception. Both verbs appear in reference to God as father begetting Jesus, but only 'cennan' is associated specifically with Mary as the mother. 'Acennan' has the additional connotation of spiritual birth in Christian contexts (*Dictionary of Old English*, *s.v.*). 'Innoð' can refer to the inside of the body: 'stomach, womb, bowels, the breast, heart' (Bosworth-Toller, *s.v.*). Given this range of definitions, it is difficult in the context of Riddle 35 to interpret 'cennan' as 'beget' rather than 'give birth to,' so it might be appropriate to think of 'wong' in this case as feminine despite its grammatical gender.

Kellie Robertson argues that in texts of the later Middle Ages, 'medieval stones were irrepressibly vital: inner virtues bestowed upon them quasi-animate powers of motion and action, while "mineral souls" linked them to the plants, animals, and humans further along the scala naturae, or ladder of nature' (92-93). She argues that in the Aberdeen Bestiary, written about 1200, 'anthropomorphizing accounts of fire-producing stones suggest a natural world motivated by recognizably human desires and behaviors' (93). The depictions of objects made from ore in the riddles also frequently feature anthropomorphizing descriptions that push the human out of, then back into, the center. Objects are described as participating in what seem like human networks, for example in the discussion of being 'born' or of contemplating heirs. But when ore calls the human the enemy, it resists affiliation with the human, locates itself potentially outside of human economies and human concerns.

Like Riddle 35, Riddle 83 is solved as 'Ore' by most commentators, though Williamson identifies it as 'Gold' (483). The speaker opens: 'Frod wæs min fromcynn' ('My origin was ancient,' ll. 1-2a). In the biblical paradigm that governed medieval thought, ore is older than humans, because it was created with the earth on the third day, and humans not until the sixth (Gen. 1: 9-31). The speaker of the riddle states that ore was made to live among men only after humans had learned to use fire, and continues: 'Nu me fah warað / eorþan broþor, se me ærest wearð / gumena to gyrne' ('Now the brother of earth remains hostile to me, he who at first did me evil among men,' ll. 4b-6a).

The identity of 'brother of earth' is vexed; Patrick J. Murphy accepts Franz Dietrich's suggestion that it is Tubalcain, 'the traditional founder of the foundry and metalworking' (142), though he points out that Williamson and Nelson take the solution to be more general (141). Murphy reads 'fah' as meaning 'cursed,' which he sees reflecting Tubalcain's descent from the cursed Cain. Another possible brother of earth is Adam, and humans in general who, as Adam's descendants, could be said to remain 'brothers of earth.' Whether the reference is to Adam, Tubalcain, or humans in general, the riddle suggests that human and stone are allied in kinship. At the same time, the speaking 'ore' places the human in a position of hostility. The idea that humans could be hostile to ore, or to earth, recorded in a riddle inscribed on parchment a thousand years ago, challenges environmentalists' easy assumptions that an awareness of the human violation of earth is a recent concern. Riddle 83 suggests that humans in general are the illegitimate usurpers of the earth's ore. The ore is centered and given temporal and ethical priority, while the human is marginalized.

Cohen argues that, taken together, the words 'inhuman' and 'nature' 'foreground the difficulties of speaking of that which is not us within narratives we fashion' (2014: ii). But Cohen also points out that various 'inhuman' objects are necessary for humans to communicate. 'Although their agency is not necessarily easy to behold, without a networked alliance of nonhumans you would not be reading these words,' Cohen writes, commenting on the word processor and the page, twenty-first century successors to the 'Book' riddle's skin, feathers, and ink. (2014: ii). Cohen further argues 'as Stacy Alaimo's trans-corporeality or what Laura Ogden describes as material entanglement make clear, segregation of human and inhuman, nature and culture belies a complicated reality, and intertwined environmentality' (2014: iv). The traces of the Anglo-Saxons survive today only in material items such as jewelry and weapons, as well as textual objects made of skins, created with the help of ore and antlers, as discussed in the previous chapter.

In considering the earth to be a 'womb' from which someone can fashion a sword that is itself barren, the riddles point to such entanglements and articulate human needs for the inhuman in a reversal of the inhuman need for the human seen in *The Ruin* (Chapter 3). Many environmentalists today argue that students and scholars must venture away from built environments to experience the 'natural' world in order to think ecologically. At the risk of essentializing an entire culture, perhaps the fact that the Anglo-Saxons lived in closer proximity to nature than most European and American environmentalists, when food and shelter were precarious and human connections to natural sources constantly foregrounded by daily living conditions, enabled them to understand human entanglements with natural objects and non-human living beings more easily than humans in developed urban areas can today in our movements among climate-controlled homes and workspaces by way of climate-controlled conveyances and our easy access to food in packages of plastic and metal rather than on the hoof or in the field.

Riddle 20 is written in the voice of a weapon made of iron, wound about with gold and decorated with precious gems. Most editors solve the riddle as 'Sword,' though Bitterli tentatively follows Trautmann's reading of 'Falcon', also suggesting 'Hawk' as a possibility (19). The subject of the riddle describes itself as a participant in human networks and economies: 'Cyning mec gyrweð / since ond seolfre ond mec on sele weorþað' ('A king adorns me with jewels and silver, and honors me in the hall,' ll. 9b-10). The sword also refers to its inability to beget children: 'Ic me wenan ne þearf / þæt me bearn wræce on bonan feore' ('I may not expect that offspring could avenge me against my killer,' ll. 17b-18). This seems to be the voice of a lifeless creature that obviously cannot bear offspring because it is not alive. Yet the example of the childless Beowulf makes it clear that the inability to engender an heir is a serious human problem. It is difficult to think in terms of the sword, an inanimate object, bearing children or being incapable of doing so. However, in the context of the 'Ore' riddles, which describe the earth as a womb, the idea of a product of the earth bearing children, or being barren, has potential significance. But Riddle 20 can simultaneously be read in terms of anthropomorphic metaphor, placing the sword literally and the riddle figuratively in service of humans. In describing miners and metalsmiths as the enemy, Riddle 83 pushes human concerns to the margins. Riddle 20 makes it possible to centralize them, but in a context that always threatens to decentralize them with different possible interpretations.

In an essay introducing a volume of Critical Inquiry devoted to theorizing things, Bill Brown limns a distinction between 'object' – a work of art or other item socially elevated to the status of museum piece, and 'thing' – an everyday item that we take for granted until it breaks or malfunctions. In Riddle 26 ('Book'), discussed in greater detail in the previous chapter, the parchment to be inscribed is a 'thing,' each leaf functionally interchangeable with other leaves. The feathers used to write, the oak gall from which tannic acid is extracted to mix with iron oxide: these are all 'things.' The book made from these things, however, is an 'object.' A book today, even a Bible, might be a mere 'thing,' identical to thousands of other copies in homes and stores and warehouses, functionally interchangeable perhaps even with millions of other printed copies. But in the year 1000, a book of scripture was a rare object, the product of numerous artisans and 'whole villages of animals' (Holsinger 619). The 'Book' of Riddle 26 is bound in gold-decorated covers, perhaps resembling the ninth-century binding of the Lindau Gospels, now in the Morgan Library.

Similarly, Riddles 48 and 59 both describe ritual 'objects' made of gold. In contrast to riddles speaking directly in the voice of its subject, these two riddles are both written from the point of view of a human viewer rather than from the point of view of the object whose identity is to be guessed. Riddle 48, 'Chalice,' begins 'Ic gefrægn ... hring' ('I have heard... [of a] ring,' l. 1), and Riddle 59 begins, 'Ic seah in healle hring gyldenne' ('I saw in the hall a golden ring,' l. 1). Each begins, then, with the riddling speaker as a human who has seen an object, which also becomes the grammatical object subordinated to the human subject. The 'ring' of each riddle is, however, immediately followed by references to humans who are also objects of the sentence, subordinated to the speaking subject 'I.' Riddle 48 continues: 'Ic gefrægn for hæleþum hring endean, / torhtne butan tungan, tila þeah he hlude / stefne ne cirmde, strongum wordum' ('I have heard of a ring deliver in strong words a message for men, splendid without a tongue, competent though he cried out with no loud voice,' ll. 1-3). This 'ring,' chalice or plate, is personified as speaking in words though it has no tongue and no voice, thus sending a strong message through its symbolic force. Riddle 59 places the 'ring' at a double remove, with the riddle's speaking voice describing the object via the men who in turn observe it: 'Ic seah in healle hring gyldenne / men sceawian, modum gleawe, / ferþþum frode' ('I saw in the hall men looking at a golden ring, wise in mind, sage in spirit,' ll. 1-3a). Riddle 48 also continues by making the 'ring' the object of the gaze of men: 'Ryne ongietan readan goldes / guman galdorcwide' ('In the red gold, men could perceive a mysterious saying, mysterious song,' ll. 6-7a). Riddle 59 echoes the emphasis in Riddle 48 on the silent symbolism of the object: 'Him torhte in gemynd / his dryhtnes naman dumba brohte' ('Splendidly, mutely, it brought god's name into their minds,' ll. 7b-8). The idea of being unable to speak yet 'torht' – 'bright, splendid, noble, beautiful' (Bosworth-Toller, *s.v.*) – appears in both riddles.

For Brown, things are important in that they are external to individual selves, and as such function analogously to living beings, including other humans. He writes: 'Theodor Adorno ... understood the alterity of things as an essentially ethical fact. Most simply put, his point is that accepting the otherness of things is the condition for accepting otherness as such' (12). In Brown's formulation, it does not really matter if the riddle is sheep or text, ore or chalice, inanimate object or animal, or even another human being. Everything is different from the perceiving human. The two riddles describing ritual objects enact this differentiation, though this is complicated by riddle 48 with its doubled objecthood, describing the object beyond the gaze of human viewers, who are also the objects of the riddle's opening 'I heard.'

Graham Harman argues from the perspective of Object-Oriented Ontology that 'objects exist at many different levels of complexity, and they are always a hidden surplus deeper than any of the relations into which they might enter' (100). He further adds, 'gold is an inscrutable object existing at countless layers of reality simultaneously' (120). This is potentially true of all objects, but Harman uses gold as an example because of its value, its limited quantities, the fact that it can be mined by humans but not created. The high monetary value of gold gives it greater economic importance in Anglo-Saxon England and greater symbolic value in Old English literature than iron. Objects made of gold are cherished not only for their economic value but also for their provenance, given for instance as gifts from queen or lord to warrior. As objects used in Christian worship, they also attain symbolic value. Interestingly, though, the 'Ore' riddles give subjectivity to the baser metal by narrating them in the first person, while the riddles describing objects made of gold do so from the perspective of human narrators.

The 'Ore' and 'Gold' riddles, then, offer a sequence of depictions that suggest different relationships with the human, sometimes making the object metaphorically human, sometimes placing the thing as the object of human vision, even doubly object of human seeing humans seeing object. Centering the object is possible, but the next line of the same riddle might push the object back to the edge with a reference to human concerns, whether literal or metaphorical, or impossible and therefore perhaps humorous – or poignant (e.g., the desire of a sword to procreate).

Like ore and the objects made from it, and like the animal of the 'Book' riddle (26) discussed in the previous chapter, trees ripped from the ground to make things for human use are depicted in the riddles as protesting their treatment. Riddle 73 is solved by most editors as 'Lance/Spear', but Trautmann proposed 'Battering-Ram' (127), a suggestive reading especially from an ecocritical perspective. The tree contrasts its life in the woods with its treatment by humans:

> Ic on wonge aweox, wunode þær mec feddon
> hruse ond heofonwolcn, oþþæt me onhwyrfdon
> gearum frodne, þa me grome wurdon,
> of þære gecynde þe ic ær cwic beheold,
> onwendan mine wisan, wegedon mec of earde,
> gedydon þæt ic sceolde wiþ gesceape minum
> on bonan willan bugan hwilum.
>
> I came forth/grew in the field, lived where ground and clouds nourished me, until in advanced years they turned me upside down, those who were hostile toward me, of that nature/kind that I earlier saw living, disturbed my being, shook me out of the earth, caused me against my nature/fate to bend indefinitely to the will of a killer. (ll. 1-7)

Anne Harris writes that, in *The Dream of the Rood*, 'the Cross does not just have consciousness, desire, and suffering: it also has memory' (34). The tree of Riddle 73 remembers its origins, much like the ore of Riddles 35 and 83 and the sheep of Riddle 26. The fact that objects are depicted as remembering their origins in different states of being is particularly intriguing in the context of the myth the Anglo-Saxons crafted of their origins on the other side of the sea, in a different land and context than that in which the *Exeter Book* riddles were written down, in the same manuscript as *Guthlac A* with its (mostly suppressed) narrative of battle by invading Saxons against native Britons (Howe 1989; also see Chapter 4). More broadly, the shared quality of possessing memory transcends boundaries among human, animal, plant, and stone, locating different kinds of being on continuums rather than on opposite sides of a series of dichotomies.

The tree has a nature, a destiny, a fate even ('gesceap,' l. 6) independent of human concerns, independent of its utility or aesthetic interest to humans. It has a will, and has been taken from the earth and bent to the will of humans, caused to act against its own will and inclinations. It is presented as having both consciousness and conscience. It is used by humans to harm other humans, a weapon in human affairs even though it calls humans the enemy. As Cohen has argued, 'things matter' (2012: 5). Stones, water, the parchment and the wood gnawed by the bookworm: whether 'things' or 'objects,' they are not simply inanimate. Cohen and Lowell Duckert argue that things have 'complicated agency,' adding that 'humans are not simply called upon to save, preserve, or conserve a lifeless material world (what hubris), but to recognize the life that already pulses within inorganic forces, manufactured and found objects, nature, and things' (2015: 3). Cohen further argues that 'things matter in a double sense: the study of animals, plants, stones, tracks, stools, and other objects can lead us to important new insights about the past and present; and ... they possess integrity, power, independence and vibrancy' (2012: 7).

Several riddles describe their suffering at the hands of human 'enemies.' Riddle 23, 'Bow,' refers to its use by a warrior as 'wite' ('torment,' l. 6). The reference to 'torment' comes between a description of the bow shooting an arrow in combat and a description of its being bent and then released to shoot the arrow. The 'torment' could refer specifically to being bent into bow-shape and then allowed to straighten, or more generally to the process of being put to human use in battle, much as the tree protests its use as a weapon in the wars between humans. Riddle 81, 'Weathervane', refers to 'wonsceaft mine' ('my misery,' l. 12) caused by a storm, which it calls 'aglac' ('monster,' l. 6) – the same word used repeatedly for Grendel and once for Beowulf himself. The shield of Riddle 5 comments on the wounds it had to suffer: 'Ic eom anhaga iserne wund, / bille gebennad, beadoweorca sæd, / ecgum werig' ('I am a lonely being, wounded by iron, wounded by sword, weary of the [sword's] edge,' ll. 1-3a).

The unusual word 'bennan' also appears in Riddle 53, in which a human narrator describes the misery of a tree turned into a battering-ram – at the hands of the humans that kill and shape it. The narrator begins, like the narrator of Riddle 59 ('Chalice'), 'Ic seah on bearwe beam hlifian' ('I saw a tree living in the wood,' l. 1). 'þæt treow wæs on wynne... oþþæt he frod dagum / on oþrum wearð aglachade / deope gedolgod' ('The tree lived in joy... until in ancient days he was deeply wounded in misery by others,' ll. 2b, 4b-6a). Though 'treow' is a neuter noun and one would expect the pronoun 'hit' ('it'), 'he' is used, suggesting perhaps that the tree is seen as masculine, perhaps that it is seen as personified and thus possessing natural, rather than merely grammatical, gender. The riddles suggest the idea of object suffering analogous to human suffering a thousand years before the possibility is contemplated again by modern theorists of the environment.

The riddles depend on polysemy, using it to trick audience members as to their solutions, but the very multiplicity of possible meanings also contributes to slippage in the boundaries between different kinds of things, alternately foregrounding and marginalizing the human. It is important that the human narrator of Riddle 53 recognizes the possibility that the tree is made miserable by human actions. The fact that three objects used in fighting – bow, shield, battering-ram – all comment on the miserable conditions of battle is also an oblique commentary on the problems that war causes for humans as well as the environment.

Trees have a particularly wide range of potential references for the Anglo-Saxons. In an often-quoted passage from Bede's *Ecclesiastical History*, the pagan chief priest Coifi accepts the spiritual superiority (or political expediency) of conversion to Christianity, gets on a horse, takes up a weapon, and desecrates the grove of trees in which he has previously worshiped, throwing a spear into it and destroying its idols. 'Ond he ða heht his geferan toweorpan ealne þone herig & þa getimbro & forbærnan' ('And then he commanded his companions to tear down that sacred place and the buildings and burn them,' II. 13, ll. 10-11).

Despite Coifi's dramatic conversion, pagan practices persisted in Anglo-Saxon England, as attested by place names, especially at the margins of political entities. In the 'Preface' to the translation of Augustine's *Soliloquies*, Alfred lists several ways in which trees can provide utility to humans, in a passage that reads like a response to lingering pagan practice:

> Gaderode me þonne kigclas, and stuþansceaftas, and lohsceaftas and hylfa to ælcum þara tola þe ic mid wircan cuðe, and bohtimbru and bolttimbru, and, to ælcum þara weorca þe ic wyrcan cuðe, þa wlitegostan treowo be þam dele ðe ic aberan meihte. Ne com ic naþer mid anre byrðene ham þe me ne lyste ealne þane wude ham brengan, gif ic hyne ealne aberan meihte. On ælcum treowo ic geseah hwæthwugu þæs þe ic æt ham beþorfte.
> I then gathered for myself cudgels, and posts, and bars and handles for each of the tools that I could work with, and wood for building for each of the jobs that I could do: the finest tree of that valley that I could carry. I did not come home with a burden, because it did not please me to bring all of the wood home, if I could carry it all. On each tree I saw something that I needed at home. (47)

Alfred's discussion places materials from the trees at a double remove from natural growth. He lists a variety of ways in which parts of trees can be used by humans, as tools as well as building materials. But his discussion of taking things from the forest for human use is simultaneously metaphorical: all the varieties of timber represent bits of knowledge to be gained from Augustine's *Soliloquies*, for which the passage stands as a preface, and perhaps also from other books belonging to Alfred's program of translation and education of his people. As Valerie Allen notes 'measurement is no self-evident exercise and ... its representations speculate as well as and maybe better than metaphor' (63). Measuring the utility of weapons made from plants in war or other human activities is a way of assigning importance to trees as objects as well as to landscape, anticipating the meditations of contemporary philosophers about the importance of things.

In 'Hewn,' Harris points out that felling trees is analogous to human death. 'The hewn becomes the inhuman. It is the thing after the cutting: wood after tree, statue after stone, jewel after gem' (19). Riddle 73 has the tree hewn down to become the battering ram that attacks the walls of human habitations. The shield of Riddle 5 protests its absorption of blows meant for humans: 'Ic eom angaha iserne wund, / bille gebennad, beadoweorca sæd, / ecgum werig' ('I am a lonely being, wounded by iron, wounded by sword, weary of the [sword's] edge,' ll. 1-3a). Echoing Alfred's *Preface*, Riddle 30, copied twice in the *Exeter Book*, describes wood as a functional substance turned to a wide variety of different human uses:

> Ic eom legbysig, lace mid winde,
> bewunden mid wuldre, wedre gesomnad,
> fus forðweges, fyre gebysgad,
> bearu blowende, byrnende gled.
> Ful oft mec gesiþas sendað æfter hondum,
> þæt mec weras ond wif wlonce cyssað.
> þonne ic mec onhæbbe, ond hi onhnigaþ to me
> monige mid miltse, þær ic monnum sceal
> ycan upcyme eadignesse.
>
> I am busy with fire, dance with the wind, wound about with glory, weather's companion, ready for a journey, afflicted by fire, blossom in a grove, a burning coal. Often fellows send me from hand to hand, so that proud men and women kiss me. Then I raise myself up, and they bow down to me, many with favor; there I shall prolong the source of men's happiness. (ll. 1-9)

The riddle describes different natural states in which trees occur, as well as various uses to which wood can be put. It can play with the wind perhaps as branches growing high or a mast for a ship; 'fus forðweges' ('ready for a journey') may reference a ship made of wood. Wood can burn, in a forest fire, to heat a home, or when the home itself burns down, as foreshadowed in *Beowulf*. Passed hand to hand and kissed by men and women, wood has been fashioned into a cup for mead, ale, or wine, as also described by several other riddles. At the conclusion of the riddle the wood is raised as a cross to be worshipped by men in order to lengthen their prosperity.

The Dream of the Rood is not a riddle, but it uses language similar to that in the 'Tree' riddles as it describes a tree wrenched from its homeland to be turned into a cross to bear Jesus during his crucifixion and then be worshiped by Christians. Much as the riddles speak in the voices of objects, the tree of the *Dream* speaks, though framed by the narration of the dreamer; it recalls being cut down and 'genaman me ðær strange feondas' ('strong enemies seized me there,' l. 30b). It is turned into a cross and, in the dreamer's vision, covered alternately with Jesus' blood and adorned with jewels. The movement from living tree to blood-stained gallows to jeweled cross echoes the movement across categories depicted in many of the riddles, notably Riddle 26 in its movement from sheep to parchment to book of scripture (see Chapter 5).

Harris ('Hewn') points out that portions of the poem *The Dream of the Rood* are inscribed in the stone of the Ruthwell Cross, enabling further shifts in the identity of the carved/bejeweled/blood-stained cross/tree/cross. 'The cross remembers itself as wood and calls itself forth as jeweled reliquary.... The ecology of the hewn is one of shifting materialities and serialized ontologies: tree becomes wood becomes Cross on stone carved with vines and animals' (27). Because the Cross speaks to a sleeping dreamer, 'human consciousness [is] eclipsed by the bright light of a dream vision' (34). Yet 'stefn' (l. 31) means not only 'root' or 'stem, trunk,' but also 'voice' – the dreamer perhaps hears the tree say not only that it has been severed from its trunk but also that it has lost its original voice as tree, so as to be able to become the cross on which Jesus was hanged.

The highly valuable materials used for decoration, the reference to the sign of the cross, and the riddling nature of the description in *The Dream of the Rood* echo the themes of Riddle 55, also about trees uprooted to make an object useful to humans. Riddle 55 describes an object made of wood and decorated with gold and jewels, bringing various literal and symbolic meanings of timber and minerals into play simultaneously. A variety of solutions have been proposed for the riddle. Tupper argued for 'Cross' (189), a solution accepted by Bitterli (129); Dietrich suggested a shield or scabbard and Trautmann a harp (Williamson 1977: 301); Williamson suggests a rack or strong-box for weapons in comparison with a rood or gallows (1977: 303); Murphy accepts 'weapon rack' (2011: 62). Muir agrees that 'some sort of sword-rack or -box seems intended... perhaps in the shape of a cross' (622) and he is followed by Niles, who argues for 'a wooden structure used to hang and/or store weapons' (2006: 75). A T-shaped or cross-shaped form for storing chain mail or other armor is another possibility, though the construction out of four species of wood suggests a decorative or spiritually meaningful object, rather than a utilitarian one.

Like the chalice/gold plate riddles, this one is narrated in human voice, rather than speaking in the voice of the object to be guessed. Like the church plate of Riddles 48 and 59, this object is placed at a double remove from the audience, seen by the narrator as it is carried by other people: ' Ic seah in healle, þær hæleð druncon, / on flet beran feower cynna, / wrætlic wudutreow' ('I saw in the hall, where heroes drink, a wondrous wooden tree brought onto the floor, of four kinds…' ll. 1-3a). It is decorated with gold, silver, jewels, 'ond rode tacn' ('and the sign of the cross,' l. 5a). The riddle goes on to catalogue the species used in making this wondrous object: 'þær wæs hlin ond acc ond se hearda iw / ond se fealwa holen' ('there was maple and oak and the hard yew and the pale holly,' ll. 9-10a). The catalogue recalls the list of different uses for wood in Riddle 30 as well as in King Alfred's 'Preface' to the *Soliloquies.* Like the 'Preface,' it merges an aesthetic description of the decorated object with awareness of the utilitarian qualities of different species of wood, and likewise seems to carry a rejection of pagan practice at the margins of Anglo-Saxon Christian and political communities.

Riddle 26, 'Book,' provides an interesting analogy: a sheep becomes parchment, which is written on by the dancing wing of a bird to become a sacred text, bound into a codex (see Chapter 5). Interestingly, in the context of the two 'Chalice' Riddles (48, 59), the book is, in the last lines of the riddle, described as decorated with gold and gems to become an object of physical value as well as spiritual significance:

> Mec siþþan wrah
> hæleð hleobordum, hyde beþenede,
> gierede mec mid golde; forþon me gliwedon
> wrætlic weorc smiþa, wire bifongen.
>
> Then a man covered me with book-covers, covered me with hide, bedecked me with gold, further adorned me with a smith's wondrous work, circled me with wire (ll. 11b-14)

The Dream of the Rood, the 'Chalice' riddles, and the 'Book' riddle all narrate transformations of materials into objects of substantial economic value as well as symbolic importance within a Christian context. The words written in the book or on the Ruthwell Cross, like the holes pierced in the tree/cross, carry meaning greater than simply their words or syntax, because they point to subjects and objects of Christian faith.

The book of scripture, the cross, the plate, and the faith they describe and symbolize, are understood as 'alive' by those of the Christian faith. The Cross and the Book 'live' in senses different from the lives of the tree and animal that enable their creation, and that metaphorical 'life' enables a different meaning than the constructed existence of Shield or Bow or Battering-Ram, all objects of human utility and more specifically implements of war associated with death. Chain Mail, Ore, and Battering Ram are described as voicing protests at being killed and transformed into items of use to humans. But these voices are anthropomorphic and metaphorical without the layer of spiritual meaning given to Chalice, Rood, and Book.

Several of the riddles describe humans as their subjects, both in the narrative voice of the person and in the narrative voice of a third-person narrator that makes the person into an object. They are not given memory, not presented in transition from one state to another, but are presented as static, much as earth and landscape are often used in literary contexts. Object-oriented ontology levels all objects, giving them the same metaphysical status. As Alan Montroso points out, Graham Harman has argued that even humans should be read as things: 'in a move quite rare for the literature of OOO, Graham Harman pauses to reflect on the nature of real objects using a human, American philosopher Richard Rorty, as an example' (Montroso 40). The *Exeter Book* riddles anticipate Harman by a thousand years when they place human bodies on the same level as objects, as in the example of the equivalence of fingers and feather in Riddle 26 (see Chapter 5) as well as in Riddle 86.

Riddle 86, whose object is a human, complicates the position of human narrator as subject and thing as object (grammatically as well as conceptually) within the riddles. Riddle 86 is narrated by a human voice that reduces the other person described to a collection of disconnected body-parts, recalling the isolated fingers of Riddle 26 (see Chapter 5). The riddle's narrator begins by calling the person 'wiht' ('creature,' l. 1), a word also used in the riddles for objects such as icebergs, in the third person. But the riddle concludes with the first-person formula, 'Saga hwæt ic hatte' ('Say what I am called,' l. 7). The being is described thus:

hæfde an eage   ond earan twa,
ond II fet,   XII hund heafda,
hrycg ond wombe   ond honda twa,
earmas ond eaxle,   anne sweoran
ond sidan twa.

had one eye, and two ears, and two feet, and 12 hundred heads; a spine and a stomach and two hands, arms and a shoulder, one neck and two sides. (ll. 3-7b)

Eyes are frequently metaphorical. Already in Old English, an 'eye' can refer not only to the human organ of vision, but also to the hole in a needle; it may also carry reference to cognitive or spiritual vision, or lack thereof. But 'an eage' in this riddle turns out to refer literally to 'one eye.' Likewise, 'earan twa': while ears as the organ of auditory intake can signify the ability to understand intellectually or to receive spiritual wisdom (*Dictionary of Old English, s.v.*), in this poem, they are simply 'two ears.' The twelve hundred heads, however, are heads of garlic, rather than human heads – heads lacking any intellectual or spiritual or sensory or emotional association. 'Heafod,' too, has multiple extended senses referring to the seat of intelligence for a human being as well as to a source, a highest point, a front (e.g., of a ship), or a beginning of something in space or in time. But the 'heafda' of Riddle 86 are simple vegetable forms that have grown underground, lacking in voice, reason, or emotion. The solution to the riddle is generally taken to be 'one-eyed seller of garlic.' It is interesting to note in this context that of all the body parts mentioned for the person in the riddle, the mouth is missing. The all-but-headless human of the riddle is as silent as the heads of garlic he is carrying; person and vegetable are comparable in their object status. The human becomes an object, in anticipation of Harman's discussion of Rorty. And the objectification of this human points to the objectification, more subtly, of other humans described in other riddles.

Gender and Ethnicity as Hyperobjects

Morton's notion of 'hyperobject' is probably applicable to the 'Storm' riddles (1, 2, and 3), describing various kinds of storm on earth and in sea, but it is even more valuable as a way of thinking about the operations of gender and its intersections with other social categories in many of the riddles. Hyperobjects, as Morton defines them, are different in scale from what humans usually are able to apprehend and can be understood only through their traces and through their relationships with other objects (1). Morton argues that it is precisely their vast size that makes hyperobjects so hard to see: 'The octopus of the hyperobject emits a cloud of ink as it withdraws from access' (39). Moreover, Morton writes, 'Hyperobjects are contradictory beasts' (47). Because hyperobjects are impossible to perceive in their entirety, they can be difficult to understand at all, and as a result they are subject to challenge by those who cannot or will not connect the various kinds of evidence for their existence. Morton argues that human beings exist inside climate change, unable to see it from an independent or objective vantage point. 'Because they so massively out-scale us, hyperobjects have magnified this weirdness of things for our inspection: things are themselves, but we can't point to them directly' (12).

In semantics, thematic roles identify both agents and causatives as the subjects of sentences that do things, but distinguish agents as sentient beings from causatives as natural forces without free will. A person is an agent, as an animal also can be; a hurricane or an earthquake is a causative. Yet Morton argues that all objects, including hyperobjects, 'forcefully exert ... the imperative' (67). They exist in large swaths in both space and time, making them more difficult to apprehend: 'We can only see pieces of hyperobjects at a time. The reason why they appear nonlocal and temporally foreshortened is precisely because of this transdimensional quality' (Morton 70).

For the weather-channel deprived Anglo-Saxons, storms are apprehensible only in their local effects, and in that context they fit Morton's definition of 'hyperobject.' The 'Storm' riddle(s), like the other riddles, are by their nature contradictory, giving hints and clues as to their solutions while also working deliberately to obscure their solutions. Hyperobjects are 'weird'; so are the subject/object solutions of the riddles, and their positioning with respect to humans. Storms and climate change are hard to access because of their size, but the riddles' solutions are deliberately withdrawn from access to those who seek to solve them, through the use of language that deliberately mystifies, confuses, and confounds.

It is not clear whether Morton thinks that, today, a 'normal' storm would count as a hyperobject, given weather imaging satellites capable of seeing storm systems in their global entirety, while predicting paths and intensities with various computer-assisted models. When he writes, 'Heavy rain is simply a local manifestation of some vast entity that I'm unable directly to see' (47-48), it seems he means a storm as impacted in its severity by global warming, not simply an 'innocent,' pre-climate-change storm. But Morton acknowledges that climate change is not the only, or the first, hyperobject, arguing that the Iranian physician and philosopher ar-Razi 'discovered hyperobjects in the tenth century.' Ar-Razi, according to Morton, points to natural disasters such as plagues or floods, and writes, 'such events create ruptures between epochs so that the time of one entire people can pass to the time of another' (66).

Writing at nearly the same time as ar-Razi, the authors and scribes of the Old English riddles evoke the notion of hyperobjects in their descriptions of storms. The first 104 lines of the *Exeter Book* riddles have been read as three different poems (Muir, Krapp-Dobbie, Bitterli) or as one long one

(Trautmann, Williamson). In either case, they describe storms through details about their natural force and their effects on human life, rather than through an overview of their origins and expected paths such as might be expected in modern weather reporting:

ic astige strong, stundum reþe,
þrymful þunie, þragum wræce
fere geond foldan, folcsalo bærne,
ræced reafige
I rise up strongly, sometimes savage, thunder mightily, sometimes cause calamity throughout earth, burn the people's houses, ravage the halls. (ll. 3-6a)

... hwælmere hlimmeð, hlude grimmeð,
streamas staþu beatað, stundum weorpaþ
on stealc hleoþa stane ond sonde,
The whale's home roars, loudly rages, waves beat the shore, sometimes throws stones and sand against the steep cliffs. (Riddle 2, ll. 5-7)

Hwilum ic þurhræse, þæt me on bæce rideð
won wægfatu, wide toþringe
lagustreama full, hwilum læte eft
slupan tosomne. ....
Swa ic þrymful þeow þragum winne,
hwilum under eorþan, hwilum yþa sceal
hean underhnigan, hwilum holm ufan
streamas styrge, hwilum stige up,
wolcnfare wrege, wide fere
swift ond swiþfeorm.
Sometimes I rush through, so that dark clouds ride on my back, widely scattering full streams of water; sometimes afterward I let them slip together... Like a powerful servant, at times I labor, sometimes under the earth, sometimes I must go underneath the high waves, sometimes a stir the waters up into a high wave, sometimes I climb up, excite the moving cloud, travel widely, swift and violent. (Riddle 3, ll. 36-39a, 67-73a)

Morton argues that climate change has taught humans ecological thinking (48), and has forced humans into a recognition of their decentered status. 'What ecological thought must do, then, is unground the human by forcing it back into the ground...' (18). But as shown repeatedly throughout this chapter and the previous one, the *Exeter Book* riddles already unground the human by describing animals and objects returning the human's gaze, anticipating Derrida, and making the human the object of comments, observations, and even direct challenges from animals, trees, and earth. The riddles also depict some humans as the decentered and object-like with respect to other humans, not only to the animals and things that are the solutions to some of the riddles.

Morton's primary concern in inventing the term 'hyperobject' is to think through social formulations around global warming as a phenomenon that cannot be seen in its entirety but can only be interpreted from signs and traces, from partial views, and thus to help to understand why people can take individual weather events, for example a snowstorm, as evidence against it. But it is also a highly useful concept for thinking about social phenomena such as 'gender' and 'race' that transcend cultures and persist across centuries. Much as Morton points out of the weather, people exist inside of gender, race, religion, dis/ability, and related social structures, and we can perceive ourselves only within them or at intersections among them, so that they are difficult if not impossible to see in their entirety.

In describing objects of everyday life, the riddles make references, sometimes oblique and sometimes direct, to human beings who use those objects. Men and women are mentioned, specifically as slaves and as Welsh but neither as English nor as free, suggesting the latter markers of status are seen as the norm, not needing characterization because they can be assumed in the absence of other descriptors. Social relationships such as the distinction between slave and free person, between 'English' and 'Welsh' or 'Briton' (which I characterize, problematically, using the term 'ethnicity'), and the conditions that structure gender are difficult to see in their entire structures and the ideologies that shape them. They are visible in individual examples. Religion can likewise be difficult to see its entirety, observable only in its symptoms and expressions. Christianity's status as the dominant religion in Europe and then the Americas for nearly two millennia makes it nearly impossible to see it as a whole, across time and space and in its myriad variations. Gender, religion, and ethnicity, like Morton's climate change and other hyperobjects, are sets of relations outside which humans cannot stand.

The descriptions of things in many of the riddles, whether 'natural' objects or things made by humans of materials such as ore, trees, and the skins or feathers of animals, are deeply bound up with notions about gender and class. As noted above, 'eorðe' is gendered feminine, and the ground and fields as a source of life are imagined as feminine even when masculine-gender words such as 'wong' are used. The ice, or iceberg, of Riddle 33 is also described as a feminine creature, one prone to violence. The solution to the riddle has been disputed. Trautmann argued that the solution could not be 'iceberg,' because icebergs were not known in the seas around the British Isles (93). Moreover, the word 'iceberg' is not attested in Old English. The lack of any record of a word in Old English is quite possibly accidental, but it is suggestive that no word for a floating mountain of ice is recorded in any surviving Middle English text, though cognates exist in Middle Dutch and Middle Low German, and 'iceberg' was apparently borrowed from Dutch only in the eighteenth century (see Bosworth-Toller, the *Thesaurus of Old English*, the *Middle English Compendium*, and the *Oxford English Dictionary*). However, Williamson notes that a traveler such as Ohthere sailed far enough north to encounter icebergs, and Anglo-Saxon sailors could have encountered occasional erratics drifting far from their sources (237-38).

Whether the solution is 'ice,' 'iceberg,' or a patch of river ice large enough to threaten a boat or ship, the object is clearly feminine: 'Wæs hio hetegrim, hilde to sæne, / biter beadoweorca; bordweallas grof, / heardhiþende' ('She was malignantly cruel, lazy to battle, bitter war-work, carved shield-walls, bold in purpose,' ll. 5-7a.) The riddle combines a narrator's description of the object with four concluding lines in the first person, enabling use of the gendered pronoun 'hio' ('she') as well as references to both 'modor' and 'dohtor' ('mother' and 'daughter,' ll. 9, 11). The identification of destructive natural forces as feminine resonates with the depiction of Grendel's mother and the mere filled with aggressive fish that attack Beowulf during his approach to her cave but disappear after she is dead.

Riddle 12 enacts a complicated series of dichotomies, including between male and female, slave and free, while also challenging boundaries between human, thing, animal, and land. It is worth quoting in full:

Fotum ic fere,   foldan slite,
grene wongas,   þenden ic gæst bere.
Gif me feorh losað,   fæste binde
swearte Wealas,   hwilum sellan men.
Hwilum ic deorum   drincan selle
beorne of bosme,   hwilum mec bryd triedeð
felawlonc fotum,   hwilum feorran broht
wonfeax Wale   wegeð ond þyð,
dol druncmennen   deorcum nihtum,
wæteð in wætre,   wyrmeð hwilum

fægre to fyre; me on fæðme sticaþ
hygegalan hond, hwyrfeð geneahhe,
swifeð me geond sweartne. Saga hwæt ic hatte,
þe ic lifgende lond reafige
ond æfter deaþe dryhtum þeowige.

While I possess spirit, I walk, slice the earth, the green fields, with my feet. If life leaves me, I bind dark Welshmen, and sometimes better men. Sometimes I give drink to dear ones, to heroes, out of my stomach; sometimes a stately wife steps on me with her feet; sometimes a darkhaired Welshwoman brought from afar bends and presses me, dumb drunkard in the dark night wets me in water, sometimes warms me pleasantly by the fire; sticks me in her lap with a wanton hand, moves rapidly, swivels me in the dark place. Say what I am called; I who living plunder the earth and serve men after death.

The verb 'swifan' makes only two other appearances in surviving Old English texts. In *Panther*, also in the *Exeter Book*, the Panther rests in a mountain cave after eating: 'ðær se þeodwiga þreonihta fæc / swifeð on swefote, slæpe gebiesgad' ('there the mighty warrior for a span of three nights twists in his dreams, busied with sleep,' ll. 38-39). The other attestation occurs in the Old English *Boethius*, in a passage about the transient nature of all things: 'Hwæt, eac se broc, þeah he swife of his rihtryne, ðonne þær micel stan wealwiende of þam heohan munte oninnan fealð' ('Look, also the brook swerves from its rightful course, when a huge stone rolling from the high mountain falls into it,' 6. 14). By Chaucer's time, the word had come to refer to sexual relations, and the *Middle English Dictionary* derives the word in that sense from Old English 'swifan.' A bawdy word for sexual activity might not be preserved in the surviving Old English corpus, with its heavy lean toward heroic and Christian poetry as well as documentary and religious prose. But this poem suggests that 'swifan' had already acquired a sexual connotation in Old English.

The boundary between human and animal is breached by the living animal's use of human speech, both before and after its death. The dead animal's skin is put to a variety of uses: leather strips used as rope for binding prisoners; a floor-covering; a bottle to hold warm water; a sex toy. As with Riddle 30 ('Wood'), the skin can be used for many thing. In life, the animal had stepped on the ground; its skin becomes, among other things, a rug stepped on by humans. In linking together stepping on the ground and killing plants, and stepping on skin from an animal already killed, this riddle also challenges the boundary between woman and animal and between woman and object. Much as the animal treads on earth, a woman treads on the animal used as a floor covering. The proud bride devolves into a 'Wale,' which seems to have meant both 'slave' and 'woman from Wales,' who seems to be using a portion of the animal's skin as a dildo. As Dorothy Yamamoto writes, 'women, despite being humans, are not accorded either the symbolic or the practical dignity of centrality within medieval culture' (10). The focus on the drunkenness and stupidity of the woman who uses the skin in its other forms deprives her of full humanity, much as does Riddle 52 (discussed below) in associating the female (grammatical) object of the riddle with the physical objects of the solution. Moreover, the woman in the riddle is given no voice, whereas the animal speaks at the beginning of the riddle, and then speaks in turn as the objects made out of the skin.

In a discussion of the film *Brokeback Mountain*, Catriona Mortimer-Sandilands and Bruce Erickson argue that dominant mid-twentieth century discourses 'attach wilderness spaces to performances of *heterosexual* masculinity' (3). They point out that relationships between humans and lived environments are constructed alongside norms, and challenges to norms, surrounding gender and sexual identity. 'The critical analysis of these locations and co-productions is what we mean by "queer ecology": there is an ongoing relationship between sex and nature that exists institutionally, discursively, scientifically, spatially, politically, poetically, and ethically, and it is our task to interrogate that relationship in order to arrive at a more nuanced and effective sexual and environmental understanding' (5). Like *Brokeback Mountain*, the *Exeter Book* riddles articulate a masculine, heteronormative context for the enactment of puzzles about things that in turn become puzzles about sexual relationships and identities. Heteronormativity is another hyperobject, as illuminated in its functions in the riddles. Men appear on horseback with hawks, a symbol of high status, or blowing horns to summon warriors, or drinking too much mead; women appear as slaves or servants, passing a cup of wine among the drinkers but not themselves drinking. In expanding on Mortimer-Sandilands' and Erickson's comments, Will Stockton notes that the 'environment' includes 'humans who are sometimes abjected as waste from the category of humanity, including nonwhites, perverts, the homeless, and the insane' (171). Unfortunately, the *Exeter Book* riddles provide several examples of that kind of abjection, as does Felix's *Vita Guthlaci* in casting native Britons as demons (see Chapter 4).

Male humans feature as the solutions to several riddles: two man with hawk riddles (19, 64), the one-eyed seller of garlic mentioned above (86), and Lot and his daughters (46). Women, however, appear within the riddles in ways that incorporate them into parts of a solution rather that as the whole of a solution. A possible exception is the one-line Riddle 76, which reads in full, 'Ic ane geseah idese sittan' ('I saw a lady, sitting alone'). But scholars have not allowed this riddle to rest. Tupper declines to give it a solution, referring to scholarship that reads it as parts of Riddle 75 or Riddle 77 (219). Williamson sees the line as forming the last line of Riddle 75, which he solves as 'piss,' with a distinction between men and women in the act of urinating (352-53); Muir agrees with the solution (687) but sees the runic line as an interpolation (669). Murphy (2011) also sees Riddle 76 as the final line of Riddle 75, but follows Niles in reading the solution as 'hound and hind' (173 n. 75). Bitterli notes that 'the capitalization and punctuation leave no doubt that ... the two entries [75 and 76] constitute two individual Riddles,' but he finds the solution to be impossible (106). Krapp and Dobbie (371) report with apparent agreement W.S. Mackie's 'possible solution' of 'hen' and are followed by Crossley-Holland (114). Any bird incubating an egg would seem a rather obvious solution. A woman involved in any of a myriad of household tasks from sewing or spinning to hulling peas would also be possible. If the vision of a houseful of children and domestic animals makes this seem improbable (though eventually, they do sleep), another possibility would be a nun in contemplative prayer. It might say more about modern scholars than about the scribe of the riddles that it has been deemed impossible to solve a riddle about a solitary seated woman. The inability to solve a riddle about a woman is evidence that gender is a hyperobject the operations of which extend from the Anglo-Saxon period into the present, as well as backward from the present, to influence the ways in which scholars continue to think about medieval literature.

Several riddles include human females in sexualized terms: examples include Riddle 12, with its reference to a woman, 'wonfeax Wale' ('dark-haired Welsh woman,' l. 8) who appears to be masturbating (Rulon-Miller); and Riddle 25 in which a 'ceorles dohtor' ('churl's daughter,' l. 6) grasps a penis/onion. In a postcolonial reading of Ælfric's narratives of Agatha and Lucy, Andrea Rossi-Reder points out that women's sexuality is an institutional concern for religion in Anglo-Saxon England because it is 'intertwined with issues of controlling marriage and reproduction' (184). Women's sexual activity in the riddles is confined to women of lower classes; from an aristocratic point of view, women's sexual freedom is less dangerous when it does not threaten the orderly inheritance of property. The presence of female sexuality in the riddles is complex, and a full treatment is beyond the scope of this chapter (and see Rulon-Miller, Salvador-Bello, D.K. Smith, and Stewart for excellent work on the subject), but some discussion has bearing on the treatment of women as objects parallel to the other objects in the riddles.

Riddle 62, which Williamson calls a 'delightful double entendre riddle,' points to the presence of sexual violence in Anglo-Saxon culture. The solution of 'penis' is ominously doubled with proposed solutions including fiery arrow, poker, and tool for boring holes, all implying that a woman is a passive recipient of male sexual energy, with varying amounts of violence implied in the different possible solutions. Solutions proposed for Riddle 61, including vagina as well as shirt, mail-shirt, or helmet, also suggest a connection between sexuality and violence toward women. The varying operations of sexual violence functioning in conjunction with heteronormativity in locating women in subordinate social positions are difficult to perceive in their connections and as a whole. As such, they contribute to the status of gender as hyperobject.

Morton argues that patriarchal discourse treats women as objects and objects like women: 'think of the gendering of cars and ships' (2013: 6). The riddles point to a similar equivalence in the gendering of ice and earth as feminine, as mothers, variously, of ore and of water and of Adam, so that the feminine is also identified insistently with motherhood (and see Dockray-Miller, *Motherhood and Mothering*). In a study of the intersections in Patristic and medieval attitudes toward Jews and women, Lisa Lampert has shown that early Christian commentators saw masculinity as the default condition, from which femininity necessarily deviates (29-32). In *De Virginitate*, Ambrose writes: 'Quae non credidit, mulier est, nam qui credit, in virum perfectum resurgit, in mensuram aetatis plenitudinis Christi' ('She who does not believe is a woman; moreover, whoever believes is raised to complete manhood, to the measure of adulthood of Christ,' *PL* 16: 270C). Some exemplary women can attain the honorary status of manhood, which is revealed as the norm. In his commentary on the third Epistle to the Ephesians, Jerome argued that women are deficient specifically as a result of bearing children: 'Quamdiu mulier partui servit et liberis, hanc habet ad virum differentiam, quam corpus ad animam. Sin autem Christo magis voluerit servire quam saeculo, mulier esse cessabit, et dicetur vir' ('As long as woman is for birth and children, she is as different from man as body is from soul. But if she wishes to serve Christ more than the world, then she will cease to be a woman and will be called man,' *PL* 26: 533B-33C). In *De Trinitate*, Augustine makes explicit a hierarchy that places woman below man: 'Caput mulieris, vir; caput viri, Christus; caput autem Christi, Deus' ('The head of woman is man; the head of man is Christ; also, the head of Christ is God,' PL 42: 930). Two manuscripts of *De Trinitate* (both now fragmentary) survive from Anglo-Saxon England. Augustine made similar comments in *De Genesi contra Manichaeos*, of which one complete manuscript from the period survives (Gneuss 57).

Aside from Riddles 63 and 80, which make passing reference to women handling vessels containing wine or mead, only Riddle 52 includes a human female described in terms that are not connected with sexual activity. Riddle 52 makes reference to a Welsh woman, apparently a slave; it is brief, and quoted here in full:

> Ic seah ræpingas in ræced fergan
> under hrof sales hearde twegen,
> þa wæron genamnan, nearwum bendum
> gefeterade fæste togædre;
> þara oþrum wæs an getenge
> wonfah Wale, seo weold hyra
> bega siþe bendum fæstra.
>
> I saw captives conveyed into the hall, two hard ones under the roof of the hall; they were of the same name, fettered firmly together with narrow bonds; near [pressing upon?] one of them was a dark-haired Welsh woman, she had power over both of their movements through fixed bonds. (ll. 1-7)

The riddle is solved variously: a pair of oxen led into a barn by a female slave, a pair of buckets tied together with rope, a flail, or a broom (Crossley-Holland 106, Tupper 185, Williamson 295, Muir 622). The stated text, in which a woman of Welsh origin is in charge of two bound criminals, interacts interestingly with the subtext, in which the same woman – of low class and possibly a slave or peasant – is engaged in agricultural manual labor, herding oxen, carrying water, or processing grain before grinding.

Commentators have always taken the 'wonfah Wale' as non-metaphorical, though all of the other elements in the riddle are subject to re-interpretation. This demonstrates the force of the hyperobject of gender across time and also shows how powerfully it intersects with the hyperobject of ethnic affiliation or, today, national status. As discussed in Chapter 4, Guthlac's body is legitimate, because he is male and Christian and of high social status; the bodies of the Britons are demonized and rendered illegitimate. The conjunction in Riddle 52 of the woman's status as both female and Welsh reduces her to an object comparable to the 'ræpingas,' ('captives'), which can be interpreted as humans or as things.

Yamamoto points out that, in medieval culture, 'all bodies are not of equal value' and that evaluating claims about which bodies are 'good' or not are complicated by the additional problem that 'individual bodies do not always stay the same' (3). She quotes Michael Camille on what he calls 'marginal art,' a category in which he includes gargoyles and marginal drawings that he finds 'expressive of resistance to the "official" culture of the primary text' (Yamamoto 5). Several of the riddles contain a text that appears to describe one thing, but behind which a different solution is hidden. In some cases, there is a marginalized 'obscene' solution as well, but in other cases, the 'marginalized' presence is the voice of ore or an antler, or of a woman of low status. Riddle 52 places a lower-class woman in power over fettered male criminals, a closely limited situation. In the various solutions, the 'criminals' come to represent animals or objects – buckets, oxen, segments of a flail – but the woman remains a woman. Her static position in the text of the riddle puts her in a position of equivalence with animals and objects, unlike the two fettered men, who stand in for objects specifically in opposition to them. The identification of a person as both woman and foreign appears to make her impervious to symbolic interpretation.

Morton suggests that feminist philosophy requires recognition of the 'presence' of things that acknowledges their continuous existence without reducing them to essential qualities. 'Ontology should respect the strangeness and uncanniness of things, acknowledging that objects are unique entities, and thus ontologically separate, no matter how much they may interact or be entangled with one another… A feminist ontology might support objects that merely exist, without any interaction whatsoever. Objects should be allowed to be inward, introverted – to exceed any gaze, any encounter at all' (Morton 2013: 65). But recognizing the presence of women as 'strange' despite entanglements with other entities does not recognize social formulations such as gender, ethnicity and class as hyperobjects. It sees the individual entities without recognizing their participation in complex networks that are apprehended only in their effects but difficult to see as a whole.

Morton's project is to develop and articulate an ethical sense that encompasses the entire world, and all the beings in it, living and non-living. 'This would constitute a move toward a democracy that included nonhumans – or better, realized that nonhumans were already part of social space, that social space was never fully human. Rather than extending humanlike powers to nonhumans, we could instead dismantle what makes humans different from nonhumans' (2013: 67). The creation of an ethics that requires that things have agency, and that places humans on a spectrum with various kinds of different non-humans, is a valuable enterprise. But like some other ecocriticisms and ecofeminisms, it does not go far enough in challenging the power of the social structures that differentiate among different kinds of humans, rendering some of them objects while others retain subjectivity and power.

Karl Steel has argued that Middle English texts use animals in opposition to humans to help define the human, but also to draw distinctions among humans. 'Medieval writers often drew on the natural subjugation of animals to degrade other humans by animalizing them and in order to generate, defend, and resist various dominant ideologies and elite practices' (2008: 4). Susan Crane argues, similarly, 'the founding human/animal dichotomy is so unstable that it has migrated all too easily *within* the human, to define as bestial certain slaves, women, colonials, criminals, and foreigners' (Crane 4). In the *Lives* of Guthlac examined in Chapter 4, Britons are cast as subhuman. Likewise, some of the *Exeter Book* riddles challenge boundaries not only between humans and animals but among humans, animals, landscapes, and objects, suggesting that the Anglo-Saxons saw some humans not merely as animals, devoid of reason, but also as comparable to inanimate things.

Riddles that articulate agency for sheep (26), tree (53), and deer (93) challenge the subordinate placement of animals and objects in a hierarchy dominated by humans, even if only briefly, and even if such provocations are elided as soon as the riddle is solved and its disruptive potential once again suppressed. The play of meanings in Riddle 52, with its competing ideas of a woman carrying buckets of water or supervising captives, similarly challenges the subjugation of the Welsh slave-woman, much as other riddles challenge the object status of animals. Bynum argues that 'marginal and disadvantaged groups in a society appropriate that society's dominant symbols and ideas in ways that revise and undercut them' (16-17). The riddle can be read to imagine a low-status woman appropriating and reimagining a dominant culture's ways of thinking about women and Welsh and slaves and simultaneously imagining an object in the process of the same kind of appropriation. A similar appropriation of dominant cultural tropes can be seen in the insistent voice of the sheep in the 'Book' riddle (26).

Rossi-Reder argues that female saints symbolize their native land: 'The violation of the woman saint's body ... reflects her occupied homeland's exploitation' (184). Ælfric makes explicit the connection between a woman's body and the landscape in his narratives of female saints when he 'compares Agatha and Lucy to components of the land – stones, rocks, and minerals' (Rossi-Reder 190). Yamamoto insists that while some discourse associates women with nature, they are not in fact therefore excluded from culture.

> In important ways, too, women are not 'outside,' or 'around' culture but profoundly involved with it. Rather than giving women a fixed place, or places, within a two-dimensional schema, I believe it is more illuminating to think of them as both wholly present and wholly absent, from the dominant, male point of view. This mixing of presence and absence generates anxiety, since men cannot always be sure which style is operative. (206-07)

Morton's suggestion that women exist as beings and not simply as assemblages of cultural operations is workable, but it risks a movement toward seeing gender as a fixed quality of a woman's body rather than a cultural artifact, or rather a vast collection of cultural artifacts impossible to apprehend in their totality, a hyperobject. Yamamoto's acknowledgement that women have a multiplicity of places within the hyperobject of culture allows a challenge to its paradigms.

Conclusion

The *Exeter Book*'s first sequence of riddles (numbers 1-59) begins with Storm and ends with Chalice. While there does not seem to be a particular order to the riddles in between those endpoints, there is a beginning in 'inhuman nature' (Cohen) and an ending in an object the description of which includes several references to God, as both creator and savior of humans. Cohen argues: 'Inhuman forces and objects ultimately refuse domestication' (2014: iv). The presence of the Chalice at the end of the first series of riddles suggests that all that has come before might be read as subordinated to a higher power. But the cultural force of the individual riddles, and the animals and objects and storms brought to life in them, does not recede thoroughly into the background despite the presence of the 'Chalice' Riddle at the end of the sequence, much as the animal that narrates its own death in the first line of Riddle 26 ('Book') is never completely subsumed into the document inscribed into its skin, and never completely disappears.

The asymmetries in depictions of male and female characters make the human a category as complex and porous as those of animals (Derrida) and things (Morton and Cohen). Humans are objectified, things metaphorically depicted as humans, and the breach in boundaries between animals and humans that has been limned by Steel and Yamamoto, and that between animals and things in several of the riddles, challenges in turn the boundaries between humans and things. As Morton writes, 'Hyperobjects seem to phase in and out of the human world: they occupy a high-dimensional *phase space* that makes them impossible to see as a whole on a regular three-dimensional human-scale basis' (70). Hyperobjects interact with more prosaic objects in the riddles, in the presentation of Storms and Icebergs by way of details about their effects rather that information about them as wholes, as well as in the formation of cultural networks within which humans can become objects, which are in turn imagined as humans, and animals challenge the boundaries among both.

John Bellamy Foster argues that ecological critique is caught in disagreements between cultural and deep ecological analysis of contemporary climate problems (and how to solve them), which then 'perpetuate the "humanity vs. nature" conceptions' rather than recognizing the human as part of (albeit alienated from) nature (18). Instead, making 'things matter,' (Cohen 2012: 7) opens the possibility of demolishing dichotomies between human and animal, soul and body, living and dead, animate and inanimate, and imagining full moral accountability and agency across entire spectrums of existence. Collectively, the Tree and Gold riddles measure the value of things as they enumerate kinds of trees and manners and materials used in decoration for human utility, aesthetic pleasure, and spiritual contemplation. Materials from the earth are given voice, and that voice provides a radical break from the usual centering of the human within human thought and expression, but at the same time the voices given to such materials are subordinated to human concerns by their context within the riddles – manuscripts written by humans, riddles created to puzzle humans, solutions designed to amuse humans. The riddles speak to audiences trying to identify their objects/subjects, through language that deliberately obfuscates and confuses in order to create the challenge and the game. The cognitive shifts between perspectives, alongside the misleading language and the descriptive shifts deny stability and conclusiveness to either perspective, pushing a re-evaluation of the place of the human and the place of things.

Alfred's meditation on the utility of trees, and Bede's narrative of Coifi's desecration of the grove, articulate a point of view that seems to suggest difficulty in taking seriously the narratives of plant and earth suffering presented in the riddles, and the transformation of sheep into scripture and tree into cross would seem to imply, within that same ideological framework, a radically different sense of importance for the religious objects that would be seen as completely 'normal' in transcending any claim to moral value for the animal or the tree. Yet the riddles also open the possibility for a reading that prioritizes the non-human, even briefly. In radically decentralizing the human, they make humans objects of a planetary subjectivity.

The metaphorical use of trees has become particularly complicated for modern philosophy. Tree metaphors and tree structures are widespread in our cultural and intellectual formulations, used to diagram sentences, schematize relationships between manuscripts, prioritize human thought in Cartesian foundationalism, or claim that Judaism has been superseded by Christianity. But these metaphors have recently been challenged. As Alfred Siewers comments, 'Hierarchy justifiably bears a bad name in the modern West, and this is at the heart of the Deleuzian critique of the arboreal in the abstract' (2014: 103). Siewers reaches back past Deleuze and Guattari to the thought of Dionysius the Areopagite to argue that trees can be rhizomatic as well, allowing for interlinked, non-hierarchical networks of thought.

The riddles describe trees and ore and sheep, high-status men with hawks and horses and an enslaved woman from Wales, yet they are still composed and written down by humans. They could be interpreted as proposing limitations in the possibilities of human voice, yet such a reading is composed, transmitted, and heard only by humans, and never by the objects or animals described within them. Cohen argues that attending to the ways in which our notions of and relationships with objects are bound up in politics can help to articulate 'a politically and ecologically engaged ethics in which the human is not the world's sole meaning-maker, and never has been' (2012: 7). Animals are different from plants, and plants are different from rocks. The examples of sea cucumbers and Venus fly traps complicate questions of how, in fact, animals are different from plants, but it is clear that there is a difference. But there is also a commonality in that both are entitled to moral consideration, and both are given an articulation of such entitlement in the riddles, even though we cannot be sure what was behind the articulation of that entitlement: a seriousness about the idea of granting moral agency to landscape, or a sense of absurdity at the idea. Even if the ascription of agency to animals and objects is suffused with laughter, the riddles still open a window to the idea.

Steel writes that critical animal theory 'describ[es] humans and nonhumans as co-constituted by their shared worlds, and ... proposes affective nonprogrammatic relations of caring, protection, and humility' (Steel 2011: 4). Cohen proposes an ethics that includes the Mississippi river and stone. Such an ethics needs also to acknowledge the fact that things, like humans, act differently and are perceived differently in different contexts. 'Relativity theory destroyed the idea of consistent objects: things that are identical with themselves as constantly present all the way down' (Morton 10). The riddles that contain double entendres can be said to do the same, but in being riddles, in making language complicated, in describing multiple phases or uses of the same object or material, they also make problematic the idea that objects are consistent.

Harman argues that objects should neither be 'undermined' to their atomic constituents nor 'overmined' to their importance to humans (106). Objects such as the plate or chalice of Riddles 48 and 59 should be reduced neither to elemental gold nor to their importance to humans in terms of great monetary value and simultaneously of religious symbol; their full range of meanings and connotations needs to be held in mind. Attending to the multiple meanings of objects has among its effects the de-centering of the human. 'Humans lose their place as the metaphysical core of the universe in object-oriented thought, but only because *no* object is allowed to occupy that core, including the inanimate sort. Instead, all objects are equally decentered' (Harman 107). In the riddles, humans are de-centered, as one object after another takes center stage as the speaking subject and object to be guessed. As demonstrated above, however, some of these objects re-center the human through descriptions that point to human presence through metaphor or by emphasizing the symbolic importance of things. Patricia Hill Collins argues, quoting Elsa Barkley Brown, that since there is no single point from which all human experience can be understood, we must constantly 'pivot the center,' remain in dialogue with one another, listen as well as speak (270-71). Humans must move out of the center not only to allow for the voices of other humans, but also to hear the voices of things and to craft an ethics that acknowledges the legitimacy of earth and ore, trees and reeds, crows and cows, alongside and equal to humans: all humans.

In the concluding chapter which follows, I attempt to bring together into a single conversation the ideas put into motion in each of the preceding chapters. I have discussed wilderness, sea, ruins, animals and objects in various configurations and in connections with several different environmentally inflected theories. Ecocriticisms make valuable claims for re-considering textual representations of the natural world and human places in it, but they are enriched when challenged by ecofeminism, postcolonial ecocriticism, critical animal studies and theories of objects. I point to some of the things that this book does not accomplish: for instance, I do not bring ecologically based criticism into dialogue with queer theory or disability study, both of which would enrich environmental cultural studies. I conclude with a brief discussion of three post-Anglo-Saxon texts as a gesture to what might be gained by reading Anglo-Saxon literary environments against those of other peoples and/or places.

7 Conclusion: Ecologies of the Past and the Future

Nearly a decade ago, I came across Cheryl Glotfelty's *Ecocriticism Reader* and realized that thinking more about Old English literary environments might make an interesting project. I began by reading the poems from the perspective of how their depictions of landscape – primarily wilderness – diverged from the descriptions of historical Anglo-Saxon landscapes in documentary texts such as the charters and the *Anglo-Saxon Chronicle* – primarily agricultural lands. I initially imagined a book about the landscapes of Old English poetry and prose. But as I began reading ecocritical theory and Old English literature against one another, I realized a wide range of additional readings was possible and, indeed, essential, including postcolonial and feminist ecocriticisms, critical animal studies, and philosophical meditations on objects and things. Rather than attempting a comprehensive study of landscapes in Old English literature, I have in this study ranged more widely. The volume that has resulted constitutes a series of investigations of different topics that fall under the broadest possible ecotheoretical umbrella, including thing theory, animal studies, ruin aesthetics, and postcolonial ecocriticism, as well as landscape and wilderness studies that investigate the liminal and barrier functions of bodies of water. In what follows, I trace some connections among the different chapters in the book and then comment on some of what the book leaves out, and propose avenues for further study. I then take a brief look at three post-Conquest texts, the *Durham* encomium, the Domesday Book, and a late entry in the *Anglo-Saxon Chronicle.* I conclude with some meditations on how literary study motivates and intersects with climate activism.

In addition to new readings of Old English texts, this project engages critically with environmental readings of literary and documentary texts. I have made the attempt to read feminist, postcolonial, and object-based ecocriticisms, as well as critical animal studies, in overlapping contexts to understand how they can challenge and enrich one another, and how reading Old English and Anglo-Latin texts can challenge them as well. Feminist ecocriticism, for instance, investigates the ways in which women are linked with nature and landscape in ways that treat both as inferior; it does not always, however, acknowledge that many men are also 'othered' and considered as 'resources' by other men. Postcolonial ecocriticism and critical animal studies, on the other hand, sometimes neglect the roles of gender. (Sheep farms are filled with gestating and nursing females, annually shorn for wool and inseminated to produce a new crop of animals; the males are almost all killed for food at a few weeks old. Similar conditions hold for the cattle in dairies.) Further study should bring disability studies, queer theory, and a fuller examination of race and ethnicity in a wider variety of contexts into the conversation.

Many, if not most, ecocritical and environmental studies ignore the medieval, even if they seek origins for environmental ideologies in the ancient world. Much ecocritical work assumes that the Industrial Revolution constituted a turning point in human relationships with the non-human. But reading across a variety of Anglo-Saxon texts, Old English and Latin as well as poetry and prose, demonstrates that assertions that humans are entitled to earthly, animal and human 'resources' are not a symptom of the Industrial Revolution, but a set of pre-existing ideologies that enabled it. More than a thousand years ago, such concepts co-existed with the suggestion that animals and trees and the ore from the earth should be given voice and agency. Both strains of ideas about humans and our relationships with animals and the earth can be found across literary and documentary texts of a variety of genres. As such, investigating pre-modern ideas and ideologies about human relationships with the non-human poses challenges and correctives to much ecocriticism.

Ecocriticisms in Dialogue

This volume has made an effort to examine a few texts from Anglo-Saxon England alongside a few ecocritical theories, and to do so in depth, rather than to provide a survey of Anglo-Saxon texts or of possible ecotheoretical engagements. As Buell wrote in 2005, any text has the potential to be read from an environmental point of view, although some have a greater surface area for ecocritical engagement than others. Each chapter has taken a different theoretical and topical focus, and I have both discussed a limited number of texts in each chapter and attempted to limit the inclusion of any given poem to one or two chapters, though *Beowulf* is so rich that I have discussed it in several different places.

Chapter 1 surveys historical and archaeological evidence about the actual characteristics of the English landscape, to provide context for what follows. In Chapter 2, on constructions of watery environments and their inhabitants, I include *Elene*, *Andreas*, *Exodus*, and *Beowulf*. Chapter 3, on human constructions that have fallen into ruin, analyzes *Exodus*, *The Ruin*, *Genesis*

A, and *Beowulf.* Chapter 4 makes a postcolonial ecocritical analysis of Felix's *Vita Guthlaci* alongside a translation into Old English prose as well as the poetic adaptation *Guthlac A* and, again, *Beowulf.* Chapter 5 investigates the animals of the *Exeter Book* riddles, without attempting an exhaustive reading, and considers Riddles 7, 8, 9, 10, 14, 24, 26, 47, 57, 77, 80, 88, and 93. I discuss the 'beasts of battle' passages of *Elene*, *Exodus*, and *Genesis A*, and *Beowulf* at some length. Finally, Chapter 6 ranges across thing theory, object-oriented ontology, and hyperobjects, finally proposing that gender and other social constructions can be usefully analyzed as hyperobjects in their objectification of certain categories of human beings. This chapter discusses a large number of Riddles: 1, 2, 3, 5, 12, 20, 23, 25, 26 (again), 30, 33, 35, 48, 52, 53, 55, 59, 62, 63, 73, 76, 80, 81, 83, and 86, as well as *The Dream of the Rood* and the Adam-Riddle of MS Cotton Tiberius A. iii. Riddle 26, 'Book,' spans both chapters, as it concerns an animal turned over the course of the text into an object.

The subjects of the chapters are linked in that they all investigate the relationships between humans and non-human entities, as well as considering how relationships with the non-human shape how humans treat one another. Throughout, I have sought to articulate an ethics of continuity and proximity rather than one of dualism and division, one that allows agency to and respects all kinds of beings, human and animal, animate and inanimate, living and non-living, microscopic and as large as the earth. I have sought to understand how literary and documentary texts illuminate continuities between Anglo-Saxon culture and our own, how carefully articulated ideas as well as unspoken assumptions about our relationships with different groups of people and with the rest of the world's beings that seem particularly modern or even post-modern are anticipated in Old English and Latin poetry and prose.

While I have attempted to keep the chapters distinctive in terms of textual and theoretical engagements, there are also various points of overlap, particularly in the overarching theme of environmental engagement with human social structures. Feminist theory as it connects with ecocriticism lies behind much of Chapter 2, with its discussion of Grendel's mother's entanglements with the mere she occupies, as well as of the absence of women in the blood-soaked community of the Mermedonians. I return to feminism in the final chapter, where I examine riddles that placed women and objects in the same planes and consider how the notion of hyperobject can help to understand gender and related social structures that claim some humans are more human than others. The notion of hyperobject could also illuminate the gender issues discussed in Chapter 2: the fact that Grendel's mother is monstrous and associated with the underground connects with larger gendered ideologies that are, like the issues discussed in Chapter 6, difficult to see in their entirety, across space and time and in different cultural instantiations.

I turn next to an analysis of the functions and constructions of ruins in Anglo-Saxon texts in Chapter 3. The ways in which humans construct and occupy dwellings is an important recent topic for ecocriticism in that it reflects environmental ideologies and social communities as well as technological ability. Dwellings and public buildings in their day-to-day use are assumed as a backdrop to the action in several Old English poems such as *Beowulf* and *Judith*, and are given occasional description. But buildings in ruin are more frequently the subject of poetry and prose. Chapter 3 engages further with *Beowulf* in a discussion of the foreshadowed ruin of Heorot in the context of an analysis of *The Ruin*, a description of a Roman building or villa in disrepair. The focus on these ruins as well as the Tower of Babel episode in *Genesis A* and the re-imagining of the sea as ruin in the Old English *Exodus* suggest an Anglo-Saxon awareness of social structures as vulnerable. The texts cast the Anglo-Saxons as the heirs to Danish, Hebrew, and Roman traditions that have in one way or another been superseded or over-ridden.

Chapter 4, on wilderness, extends the work done in Chapters 2 and 3 on the sea and on ruins. Wilderness is cast as untracked space, much as the sea is constructed in medieval as well as modern thinking, as discussed in Chapter 2 which includes a discussion of both open sea, in conjunction with sea crossings in *Andreas*, *Elene*, and *Exodus*, as well as of marshland. Sea and wilderness landscape are both populated by animals, however, which are generally ignored. Wilderness in modern contexts, particularly in North America, has often been 'created' out of land actually occupied by farmers or Native peoples, who are forced to move elsewhere. As discussed in Chapter 3, the prose Guthlac *Vitae* describe the marshland in which Guthlac makes his hermitage, a liminal space like Grendel's mother's mere that is neither land nor sea. However, the poetic *Guthlac A* reimagines Guthlac's home as a dry, mountainous landscape, perhaps following the descriptions of the wilderness in biblical poems like *Genesis A* or in lives of desert saints.

Sea-kennings like 'whale's road' and 'swan's path' suggest that the Anglo-Saxons believed the sea to be populated as the legitimate home of non-human creatures, but many other textual moments depict it as empty space, usable as a human resource. Thinking through the agency of animals as well as 'othered' humans links representations of wilderness and sea as unpopulated by beings that matter. The *Vita Guthlaci* and the Old English versions convey this attitude toward the prior inhabitants of the marshland where Guthlac makes his hermitage, calling it trackless even as it describes the traces of prior habitation. Guthlac as warrior against actual Britons is folded together with Guthlac as spiritual warrior fending off 'demons' who speak in the language of the Britons, thus demonizing the actual prior human inhabitants of the territories colonized by Angles and Saxons.

The space Guthlac chooses to build his hermitage is, in fact, a ruin, a place previously used as both grave and cistern, suggesting different kinds of human utility and occupation. It also shows evidence of destruction by grave-robbers seeking treasure that might have been interred with the body. The crumbling structures of *The Ruin* and the unfinished Tower of Babel suffer from the absence of humans who can complete construction or make repairs. In the absence of humans they become a kind of wilderness. While ruins are clearly not 'untracked,' the landscape of Anglo-Saxon England was dotted with Roman remains which the new inhabitants lacked the knowledge to repair and which were therefore of little use unless a sarcophagus could be reused or a stone slab repurposed as a bridge. The ruined space of the island that Guthlac occupies is explicitly and redundantly described as wilderness in the Latin *Vita* and the Old English prose translation as well as in *Guthlac A*. Centuries later, European colonists would appropriate non-European lands, call them 'wilderness,' and kill or enslave their native inhabitants. The ideology that enables that much later colonizing violence is articulated in *Guthlac*'s construction of ruin as wilderness.

Women are almost completely absent from the versions of *Guthlac*, though there is reference to 'rapinas,' which encompasses plunder as well as rape. Seizing women as prisoners and raping them is an apparent staple of Anglo-Saxon warfare, as it has been in many other cultures and eras. While I have not make an ecofeminist reading of the Guthlac *Vitae*, the identification of liminal marshland with the monstrous feminine, and Felix's easy assumption that Guthlac holds dominion over the land as well as its nonhuman creatures, sentient and non-sentient alike, occurs within a context that associates masculinity with reason and humanness and femininity with nature and a lack of reasoning capacity. Cultural context, including the Latin tradition of commentary on the role of women and the association of women with monstrosity, would enable such an argument, rather than direct textual evidence within the Guthlac texts.

Chapter 5 turns to a discussion of animals in Old English texts, primarily focusing on the *Exeter Book* riddles but also examining the slaughter of animals and the claims that sea creatures are monstrous in *Beowulf* as well as the metaphorical use of wolf, raven and eagle in 'beasts of battle' passages in *Elene*, *Exodus*, and *Genesis A*. While the epic poems instrumentalize animals as metaphorical for human concerns or disposable in a demonstration of Beowulf's masculine prowess, several of the riddles describe wild animals of no utility (or danger) to humans. Moreover, many of the riddles are written in the animals' first-person voices and even refer to humans as 'the enemy.' The riddles suggest that the Anglo-Saxons understood that animals existed outside of human needs or interests, and even that animals possessed agency. How people understand their relationships with animals contributes to their understanding of themselves as human, as well as of some humans as less human than others. Reading the animals of the *Exeter Book* riddles in the context of *Beowulf* and the Guthlac narratives demonstrates a broad and variable understanding of animals in Anglo-Saxon culture. Some threads of discourse anticipate colonizing violence in making animals and some humans objects for human use. Yet other threads demonstrate a different perspective in early English culture, one that saw animals as interesting for their own sake.

Chapter 6 extends the analysis of the *Exeter Book* riddles, focusing on those that describe objects (though Riddle 26, about a sheep that is made into a book, inscribed with scriptural or liturgical texts and bound in gold, is discussed in both chapters). Extending animal rights activists' arguments, thing theorists and object-oriented ontologists argue that things, too, should be understood as having agency. Some of the riddles give voice to objects that say they have been torn from the earth by human enemies before being made into weapons or armor. While the point of riddles is to describe familiar things in strange ways, and the voicing of animals and objects is done in the service of fooling the audience, these riddles still open a space for seeing animals and objects alike as intrinsically worthy of consideration. As noted above, I conclude the chapter with a discussion of the hyperobject, a word coined by Timothy Morton to discuss climate change, and extend its range of reference to social constructions such as gender and ethnicity, returning to and connecting with the gender constructions examined in Chapter 2.

Some Proposals for Future Research

The structure of the volume leaves a great deal out. *Beowulf* and the *Exeter Book* riddles could both sustain full book-length studies on their own, as could readings of Old English and Anglo-Latin texts focused on post-colonial ecocriticism, ecofeminism, or critical animal studies. I have analyzed the

Guthlac *Vitae*, *Beowulf*, and the *Exeter Book* riddles at length, and have included less thorough discussions of portions of *Elene*, *Andreas*, *Genesis A*, and *Exodus* as well as some shorter poems; I have given short shrift to *The Seafarer* and *The Wanderer*, poems that might seem obvious candidates for ecocritical analysis alongside, for example, *Panther* and *Whale*. I have read some Latin texts against Old English translations and adaptations, notably the *Vita Guthlaci* but also passages from the Vulgate Genesis and Exodus. It would be instructive to read *Elene* and *Andreas* against their Latin sources and analogues, as well as with attention to other Old English versions of the narratives; a reading of additional Anglo-Latin materials would surely also be valuable. A reading of the *Exeter Book* riddles in the context of riddling passages in *Adrian and Ritheus* and the *Solomon and Saturn* dialogues, as well as of Aldhelm's *Aenigmata*, would also be of value.

The introduction to this volume makes a brief survey of the ecocritical avenues and theories that have been of value in the book as a whole, and defines some of the terms used by ecocritics, noting that such terminology does not correspond very well to the words found in surviving Anglo-Saxon texts. I have not attempted a full analysis of the vocabulary related to animals and the natural world. A thorough lexical study of such terminology in comparison to the words used and their ranges of reference in Latin and in Old Norse, among other cognate languages, would open interesting avenues for understanding the etymologies of current vocabularies of the environment.

The range of depictions of the sea in Old English poems is broader and more complicated than the range of images of rural landscapes, wilderness or cities. The poetry includes playful images of the sea during an enjoyable sea-journey as well as depictions of the sea as a hard, cold, and difficult place; it also sometimes imagines the sea as an instrument of God to protect or punish human beings, in terms of its watery actuality as well as in terms that make of it a fortified dwelling. The complexity of depictions of the sea is perhaps due to the fact that it is, in Old English poetry, constructed by traditions of both Latin texts and Germanic oral formulaic verse as well as representing ways in which Old English poets had actually experienced it. I have given some consideration to direct Latin sources for Old English poems that mention the sea, but have not explored larger cultural formulations of seascape in the early medieval Latin traditions. The recent volume on *The Maritime World of the Anglo-Saxons* demonstrates that many other fruitful avenues of study remain. Marshlands as sites of liminality and the limits of the human, and the constructions of (sometimes feminine) monstrosity in opposition to masculine human norms, could also be studied in greater detail and with attention to both literary and documentary texts from the Anglo-Saxon period within ecological contexts.

The study of ruins raises the interesting possibility of thinking about ruined bodies, in terms of decapitation and dismemberment as well as in conjunction with studies of disability and illness. The mutilated body of the dragon is thrown into the sea in *Beowulf.* In *Judith*, a bag used to carry cheese is repurposed to sneak the head of Holofernes back to the Hebrew city of Bethulia. In *Andreas*, Matthew's body is transformed into that of an animal, but he retains his human mind; Andrew is dragged over rocks until he is bloody, but then miraculously healed. Documentary texts describe various illnesses; famously, King Alfred suffers on-going stomach troubles, though they do not seem to interfere with his capacity to lead negotiations and armies. I have discussed ruins, but not non-ruined buildings. A literary and historical study of how the Anglo-Saxons lived, how their buildings interacted with social and political structures as well as their ideas about the environment, would be of value. The functions of different kinds of dwellings in *Andreas* and *Judith* as well as the metaphorical use of wood for construction and buildings in the *Ecclesiastical History* as well as in Alfred's *Preface* to the translation of Augustine's *Soliloquies* would reward further consideration.

While I have read the Guthlac narratives with attention to post-colonial ecocriticism, I have not attended to *Beowulf* as a narrative embedded within colonization. *Beowulf* is also a poem with colonizing meanings in its language and transmission. It is a poem about Danes, written in English several hundred years after the events it narrates, several hundred years after those Danes colonized England. The people who composed and recorded the poem were the descendants, culturally if not directly, of the people who settled England, displacing the language and customs of the native Britons, if not the people themselves. Seth Lerer sees the description of the stone floor of Heorot as 'a remnant of Roman colonization of Britain' and Hrothgar and his people also as colonizers of a landscape foreign to them: 'The *pagus* on which Hrothgar builds his hall is already inhabited. There Grendel lurks' (87). It would be of considerable interest to undertake a more sustained reading of the Grendel-kin in comparison to the British demons that harass Guthlac. *The Fight at Finnsburh*, Bede's *Ecclesiastical History*, and *Judith* are among other Anglo-Saxon texts that would reward post-colonial ecocritical readings.

I examine the reconfiguration of Guthlac's hermitage, historically fenland, into an apparently mountainous desert in *Guthlac A*, in the context of descriptions of wilderness in *Genesis A*. I have given very little attention to Latin saints' lives, and their constructions of wilderness, that could have influenced Felix or the Old English translator or versifier. The largely imagined landscapes of wilderness in Anglo-Saxon literature poses a challenge to ecocritical discourse as it has played out to date. Ecocritical thinkers have observed that wilderness itself is a human construction, conceived of in opposition to the perception of urban areas as squalid environments whose very existence comes at a great cost to the health of the environment. Observing the relationships between wilderness and urban environments as imagined in Anglo-Saxon texts allows a focus on the constructed nature of wilderness as well as of cities, then as now.

Critical Animal Studies is a robust field, with a body of scholarship and theoretical engagement with some points of commonality with ecocriticism but which has developed largely independently. 'From a CAS perspective, the twenty-first century represents a pivotal period in which ecology and animal life face unprecedented threats. In this sense "critical" expresses the urgency of our times in the context of ecological crisis' (Taylor-Twine 2). The intensity of interest in animals in medieval texts suggests that these concerns are not new. Karl Steel and Susan Crane have written excellent books about the intersections and points of challenge between medieval literature and Critical Animal Studies. In this chapter, I take a necessarily brief and somewhat preliminary look at how the field can reshape or allow for new readings of some Old English poems. As with other chapters, this topic could fill a monograph on its own, and I have only sketched out some ideas about the topic. Animals feature, sometimes prominently, in numerous Old English poems, homilies, and documentary and liturgical texts, and a full consideration would be another valuable study.

I extend Morton's concept of the hyperobject to encompass gender and other social constructions through the objects of the *Exeter Book* riddles. An analysis of the workings of gender as hyperobjects in poems such as *Genesis A*, *Judith*, and *Elene* could reveal new insights about those poems, as might a discussion of Guthlac materials that considered religion and cultural identity as hyperobjects. In the same chapter I consider objects only as found in the *Exeter Book* riddles – not any of the actual objects of the Anglo-Saxon archaeological record, or the objects described in documentary or liturgical texts, or depicted in use in other poems. A fuller treatment of materiality in Old English texts would be of interest to the field, allowing for new readings of the poems and providing new ways of thinking about the relationships between text and object. In a post-colonial reading of *Beowulf*, Jonathan Wilcox examines objects in the poem as revealing important social and cultural issues. 'Hoard and poem both tell a story which masks as much as reveals the significant structures of the day' (159). Other poems that contain important objects include *The Dream of the Rood*, to which I have given some consideration, primarily through Anne Harris's excellent reading. An investigation of the cross on which Jesus was supposed to have been crucified, as well as the nails, and their functions in the depiction of the Jewish other and the conversion of the interestingly named Judas would be a fascinating contribution to an Anglo-Saxon Object-Oriented Ontology. The contribution of constructed environments to constructions of ability and disability is a relatively recent area of study. The intersections of literary ecocriticism with disability studies would also reward exploration.

After the Anglo-Saxons

Reading Old English texts in broader medieval contexts, temporally and/or geographically, with attention to continuities as well as ruptures, would be of interest. I have limited the scope of this volume to the period prior to the Battle of Hastings, even though the Old English language continued in use for at least another few generations, and scholars such as Elaine Treharne have put the point of transition to Middle English as late as the early thirteenth century.

Norman rulers who came to England following the Conquest seem to have held ideas about landscape somewhat different from those of the Anglo-Saxon aristocracy, as suggested by the encomium poem *Durham*, celebrating the living city in contrast to the lament for the past of *The Ruin*; the *Domesday Book*'s unprecedented documentation of the lands of Anglo-Saxon England and the numbers of animals and humans they could support, suggesting an equivalence between (English) people and livestock; and the post-Conquest reference in the *Anglo-Saxon Chronicle* to the negative effects of urban construction. The colonizing actions of the Anglo-Saxons, as documented in the *Vita Guthlaci* and its Old English translations and adaptations, are echoed in the Norman appropriation several centuries later of territories and that had become 'English.' The apparent shifts in environmental imaginations suggest that a study of continuities and ruptures in Anglo-Saxon and Norman ecological ideologies would be a valuable endeavor. Here, I will briefly trace some of the early indicators of such changes in attitudes.

In his student edition, *A Choice of Anglo-Saxon Verse*, Richard Hamer prints *Durham* immediately after *The Ruin*, suggesting the two pieces frame the Anglo-Saxon period. D. R. Howlett sees the two poems as extremely similar in genre and structure (291). Differences in diction suggest that the author of *Durham* did not know *The Ruin*, but structural similarities suggest that both poets drew on 'common models' of 'the ancient genre *encomium urbis*' (293). Hamer agrees with several other scholars on a long-accepted date of composition for *Durham* between 1104 and 1109, but recently that dating has been challenged. Joseph Grossi accepts the usual dating between 1104 and 1109, and reads the poem as an articulation of the concerns of the community at that time, while Seth Lerer, also accepting the usual dating, points out that the poem has been read both as 'an eloquent survival of traditional techniques of verse-making' and 'an antiquarian *tour de force* re-creating for a literate audience the older forms of poetry for purposes politically and culturally nostalgic' (8). Peter Evan argues for extending the *terminus post quem* for the poem to 1115, based on historical, documentary and linguistic evidence.

Thomas O'Donnell, conversely, argues for a much earlier date between 1050 and 1083 based in part on the archaic alliteration and meter, which other scholars have taken as either a survival of earlier forms or a deliberate archaization. He acknowledges that the phonology of the poem's unstressed syllables suggests a late date, though he suggests that this might be an artifact of transmission rather than evidence of late composition. O'Donnell points out that Cuthbert's remains had already been in Durham by 995, and the reference to their presence does not require the assumption that they had already been translated to the new cathedral. Finally, he imagines a secret translation of Bede's remains, covered up by Symeon after 1104, claiming, 'there is nothing far-fetched about this scenario' (138).

From an ecological point of view, it is the not the generic similarities, but the differences in detail between *The Ruin* and *Durham*, that are of greatest interest. *Durham* describes a flourishing city holding the relics of several important Anglo-Saxon saints, suggesting the presence of people who worship them and maintain their crypts, while *The Ruin* depicts a building or a city in disrepair, its builders and repairers long gone. While *The Ruin* does not mention flora or fauna, *Durham* describes a world that includes animals as well as features of the landscape:

Weor ymbeornad,
ea yðum stronge, and ðer inne wunað
feola fisca kyn on floda gemonge.
And ðær gewexen is wudafæstern micel;
wuniad in ðem wycum wilda deor monige,
in deope dalum deora ungerim.

> The Wear runs around it, river strong with waves, and there in the company of the floods live many kinds of fish. And is it grown up with a great, securing forest; in those places, many wild animals live, innumerable animals among the deep dales. (ll. 3b-8)

Other earlier Old English descriptions of dwelling, such as Heorot in *Beowulf*, Bethulia in *Judith*, or Mermedonia in *Andreas*, do not include details about vegetation; where there are animals, these are beasts of battle or other wild creatures. The descriptions of the natural world in the *Exeter Book* riddles occur outside of the context of any reference to dwellings or urban localities. The non-urban environment of *Andreas* contains mainly the rocks against which the demons rack Andrew's body, leaving him bruised and bloody, while the non-human world in *Beowulf* is described as distinct from and in contrast to depictions of dwelling. *Durham* also makes the interesting claim that 'Eardiæð æt ðem eadige in in ðem minstre / unarimeda reliquia' ('Living with the blessed one in the monastery are countless reliquaries,' ll. 18-19). The innumerable reliquaries 'living' with the tomb of Cuthbert in the Durham cathedral are ruins, fragments of once living bodies of saints, while the body of Cuthbert is an un-ruin or perhaps anti-ruin: a body that, it is claimed, does not decompose.

Interestingly enough, the *Domesday Book* contains no entry for Durham, though it does catalogue nearly all of the cities and towns in England, in the first attestation of a survey of landscapes with a calculation of the value of each plot of land based on how many animals, crops and humans it could support. There are several entries for the towns surrounding Crowland, the site of Guthlac's hermitage and later of an abbey. For instance:

> In BUCKNALL, Gamal had 10 bovates of land to the geld. [There is] land for 10 oxen. In the same place [is] SOKELAND of Belchford, 10 bovates of land to the geld. [There is] land for 10 oxen. Now St Guthlac has 1 plough there in demesne; and 5 villans and 2 bordars and 8 sokemen having 1 plough. There are 120 acres of meadow, and 50 acres of woodland pasture, and 70 acres of scrubland. TRE worth 30s; now the same. Thorald the sheriff gave this land to St Guthlac for his soul (trans. Williams and Martin).

Fields for crops and pasture have no absolute size, but are defined in terms of their utility. A 'bovate' is an area of land that can be plowed by a single ox. This particular village supports fifteen men and, presumably, their families: eight 'sokemen' or freemen, two 'bordars' or cottagers and five 'villans,' of higher status than 'bordars,' but subject to the lord of a manor.

In 'TRE' or 'Tempore Regis Edward,' i.e. according to the values current at the time of King Edward, the value of the village was 30 shillings. Oxen, plows, meadow, woodland and people are all considered together under an assessed monetary value. The *Exeter Book* riddles blur categories of human, animal, and object (see Chapters 5 and 6), but the *Domesday Book* goes farther in reducing animals, humans, and landscape to their cash value and potential for generating taxes.

A post-conquest entry for 1097 in the *Anglo-Saxon Chronicle* is the first to reference urban construction, and it does so in a way that describes damage to rural environments as a result of the king's various urban building projects:

> Eac manege sciran þe mid weorce to Lundenne belumpon. wurdon þærle gedrehte. þurh þone weall þe hi worhton onbutan þone tur. & þurh þa brycge þe for neah eall toflotan wæs. & þurh þæs cynges healle geweorc þe man on Westmynstre worhte. & mænige man þær mid gedrehte
> Also many shires that were concerned with construction in London were harshly afflicted through the wall that they built around the Tower, and through the bridge that was nearly carried away by flood, and through the work on the king's hall that they built in Westminster, and many men were afflicted with that. (*Two Saxon Chronicles* 234)

The brevity of the chronicle entry leaves it unclear whether the burden to the shires occurred because materials were removed for construction or because the men who were conscripted for the construction were taken away from the agrarian labor essential for the harvest. The entry focuses upon the vexing affliction to the workmen themselves, perhaps on account of being asked to do hard labor by their conquerors, and the passage undoubtedly is politically charged. From an ecocritical perspective, it is of particular interest to note that urban construction is described as having a negative consequence on the hinterlands.

While earlier *Chronicle* entries note the deleterious effects of battle, particularly in slaughter of cattle and burning of fields, this is the first entry to observe that construction also has a cost. The description of the building of Heorot notes no negative side effects of construction, suggesting instead that it is a boon to those from far and wide who contribute materials and labor. The *Chronicle* entry for 1087, alongside the catalogue of human, beast, and land values in the *Domesday Book* and the encomium for *Durham* together suggest a change in ideologies about landscape and the humans, animals, and vegetation that inhabit it. But the ways in which environmental attitudes did or did not evolve from the Anglo-Saxon through Norman periods is another area that warrants further study.

Ecocritical Ethics and Activist Scholarship

I take for granted that we are now living in the Anthropocene, the ecological age dominated by the human for which starting points have been proposed ranging from the start of farming to the Industrial Revolution or the detonation of the first atomic bomb. Human intervention in environmental phenomena has a long history, but recent years have seen an acceleration of climate change which is, in turn, drastically affecting the contours of lakes and coastlines and individual living conditions as well as political conditions worldwide. The scholarship which underlies this volume is motivated by my conviction that climate change is real and is a great threat to global stability today. People in developed nations need to make major, perhaps drastic, changes in our living conditions and our assumptions about the availability of 'natural resources' in order to mitigate major planetary impacts already in process.

Some scholars argue that personal commitments should be excised from scholarly work. For me, the very act of engaging in environmentally committed research is itself an act of engagement with politics – a statement that these ideas are important enough to devote years of my life to investigating them. As Timo Müller argues 'We perceive nature not through language but within it' (74). Investigating the language used for natural phenomena a thousand years ago helps to think critically about the ways in which humans who have inherited the language of the Anglo-Saxons, albeit in mutated form, along with many of its cultural formations, engage with the natural world today. Though Timothy Morton suggest that climate change is incomprehensible, Scott Slovic argues for the importance of narrative in understanding the statistics about global temperatures and increasingly violent storms (27-28). Examining how people of the distant past told stories about their environments can help us to understand contemporary stories.

The study of literature is not merely an exercise in historical fact-finding or of excavating continuities and/or distinctions between past and present, but also a matter of ethics. This point is essential to an activist ecological commitment: we cannot seek to save the planet purely for the sake of the humans occupying it, but we must seek to ameliorate climate change, species extinction, and environmental degradation for the sake of the planet itself and all of its beings, sentient or not. We cannot 'save the planet,' or even 'save the whales,' unless we recognize the deeply intertwined discourses that assume and therefore create dichotomies between different kinds of beings, whether between different kinds of humans, between humans and animals, between sentient and non-sentient beings, and more. Literary analysis needs to be more than just words. It must be a call to action and activism.

Behind and beside this project lies the effort to develop and articulate an environmental, ecotheoretical ethics that would govern my own relationships with the world and could encourage others to change their own habits. Such an ethical stance requires, for me, a rejection of dichotomies and a recognition that all beings exist on a continuum where boundaries are difficult to apprehend because genders and ethnicities and religious affiliations and animals and things shade into one another and collide with one another and enrich each others' being. As Leo Mellor argues:

> This question of … the unalterable 'otherness' of the creature engaged with, has become one of significant philosophical import in ecocriticism – including how far poetry might then be able to shape a language of limit, of understanding the gap as *itself* a thing of beauty (110).

Speaking for the earth and its creatures does not merely raise interesting problems. It is crucial to fighting climate change.

But such an ethic also requires a recognition that as a species, humans (whatever 'human' might mean) need to reverse course, re-engineer our relationships to stuff and to one another, and re-think our ways of being in the world. The planet is beyond 'sustainability.' We need to recognize that maintaining current levels of consumption has us on a fast track to global catastrophe. Jonathan Maskit argues that environmental philosophers need to give more attention to the problems of excess consumption; he points out that proposals to limit environmental degradation by advocating green consumption, seeking more efficient means of production, or advocating government policies to get someone else to reduce consumption, for instance through carbon taxes on production, miss the point. 'Environmentalism might then be an opportunity to rethink who we are as subjects and how we want to live rather than a set of practices often perceived to be ascetic' (14). As Timothy Morton cogently asks 'What exactly are we sustaining when we talk about sustainability? An intrinsically out-of-control system that sucks in grey goo at one end and pushes out grey value at the other' (2013: 113). Conceiving of 'sustainability' as the goal of an environmental ethics misses the point: our growth-oriented ideologies and practices of ever-increasing consumption are inimical to any kind of ecological equilibrium, and those of us who live in developed countries need to make drastic changes at individual and collective levels as soon as possible. Yves Abrioux goes further, locating environmental degradation in the basic tenets of European and American thought. 'It is necessary to set aside the ethnocentric triumphalism of Western imperialism, which reaches deeply into contemporary thought, in order to appreciate the existence of ecologies which regulate relations between humans and nonhumans on lines other than those of subject and object' (253).

We need to reinvent our interactions between human life and the earth and all of its entities, living and not, sentient and not. Morton argues: 'What we need is an ethics of the other, an ethics based on the proximity of the stranger' (124). We need an ethics that compels us to care – in the sense of caring about, but also in the sense of caring for – people and things and landscapes that we do not personally need. Finally, and crucially, I see environmentally inflected literary study as necessarily possessing an activist component. We need to act on climate change, individually and collectively. We need to divest from fossil fuels, invest in public transit, seek ways to sequester the carbon we have released by burning things, and stop consuming so much stuff. These are not just words. This is a call to action. Stop. Think. Reverse course.

Works Cited

Primary Sources

Ælfric's Anglo-Saxon Version of Alcuini Interrogationes Sigeuulfi Presbyteri in Genesin, edited by G.E. MacLean. Halle: E. Karras, 1883.

Ælfric's Catholic Homilies: The First Series, edited by Peter Clemoes. Early English Text Society supplemental series 17. Oxford: Oxford University Press, 1997.

Ælfric's Colloquy, edited by G.N. Garmonsway. London: Methuen, 1966.

Ælfric's Lives of Saints, edited Walter W. Skeat. Early English Text Society original series 76, 82. London, 1881, 1885 [1966, 1999].

Anglo-Saxon Version of the Life of St. Guthlac, Hermit of Crowland, edited and translated by Charles Wycliffe Godwin. London: Smith, 1848.

Bede. *Historia Ecclesiastica Gentis Anglorum*, edited by Bertram Colgrave and R.A.B. Mynors. Oxford: Oxford University Press, 1991.

Beowulf, edited by Friedrich Klaeber. 3rd edition. Lexington: D.C. Heath, 1950.

Cynewulf's 'Elene', edited by P.O.E. Gradon. Exeter: Exeter University Press, 1996.

The Domesday Book: A Complete Translation, edited by Ann Williams and Geoffrey H. Martin. London: Penguin, 2003.

Electronic Sawyer: Online Catalogue of Anglo-Saxon Charters, edited by Simon Keynes and others. London: King's College London, 2016. Accessed 1 August 2016. www.esawyer.org.uk.

Epistola Alexandri ad Aristotelem, edited by W.W. Boer, Beiträge zur klassischen Philologie 50. Meisenheim am Glan: Hain, 1973.

Exeter Anthology of Old English Poetry, edited by Bernard J. Muir. Exeter: University of Exeter Press, 1994.

Exeter Book, edited by George Philip Krapp and Elliott Van Kirk Dobbie. *Anglo-Saxon Poetic Records* 3. New York: Columbia University Press, 1936.

Exeter Book of Riddles, edited and translated by Kevin Crossley-Holland. New York: Penguin, 1965.

Exodus, edited by Peter J. Lucas. Exeter: Exeter University Press, 1994.

Felix, *Life of Saint Guthlac*, edited and translated by Bertram Colgrave. Cambridge: Cambridge University Press, 1956.

Jerome. *Biblia Sacra*. In *Patrologia Latina*, edited by J.P. Migne, vols. 28, 29. Paris: Migne, 1845, 1845.

Junius Manuscript, edited by George Phillip Krapp. In *Anglo-Saxon Poetic Records* 1. New York: Columbia University Press, 1931.

King Alfred's Old English version of Boethius De consolatione philosophiae, edited by Walter J. Sedgefield. London: Clarendon Press, 1899.

King Alfred's Version of Augustine's Soliloquies, edited by Thomas A. Carnicelli. Cambridge: Harvard University Press, 1969.

Klaeber's Beowulf and the Fight at Finnsburg, edited by R.D. Fulk, Robert E. Bjork, and John D. Niles, 4th edition. Toronto: University of Toronto Press, 2008.

Lateinischen Bearbeitungen der Acta Andreae et Matthiae, edited by Franz Blatt. Giessen: Töpelmann, 1930.

Laws of the Earliest English Kings, edited and translated by F.L. Attenborough. Cambridge: Cambridge University Press, 1922.

'Lindau Gospels' (manuscript), New York: The Morgan Library and Museum. Accessed 20 July 2016. http://www.themorgan.org/collection/lindau-gospels.

Old English Boethius: An Edition of the Old English Versions of Boethius's 'De Consolatione Philosophiae', edited and translated by Malcolm Godden and Susan Irvine. 2 vols. Oxford: Oxford University Press, 2009.

Old English Riddles of the Exeter Book, edited by Craig Williamson. Chapel Hill: University of North Carolina Press, 1977.

Old English Version of Bede's Ecclesiastical History of the English People, edited by Thomas Miller. Early English Text Society original series 96. Woodbridge: Boydell & Brewer, 1997 [1891].

Patrologia Cursus Completus [Patrologia Latina], edited by J.P. Migne. Paris: Migne, 1841-1855.

Riddles of the Exeter Book, edited by Frederick Tupper. Boston: Ginn/Athenæum, 1910.

Rule of St. Benet: Latin and Anglo-Saxon Interlinear Version, edited by H. Logeman. Early English Text Society original series 90. Millwood: Kraus Reprints, 1981 [1888].

Sources and Analogues of Old English Poetry: The Major Latin Texts in Translation, edited and translated by Michael J.B. Allen and Daniel G. Calder. Cambridge: D.S. Brewer, 1976.

Two Saxon Chronicles Parallel, edited by John Earle and Charles Plummer. 2 vols. Oxford: Clarendon Press, 1892 [2000].

Vercelli Book, edited by George Phillip Krapp. In *Anglo-Saxon Poetic Records* 2. New York: Columbia University Press, 1932.

Vercelli Homilies and Related Texts, edited by D.G. Scragg. Early English Texts Society original series 300. Oxford: Oxford University Press, 1992.

Secondary Sources

Abrioux, Yves. 2009. 'Intensive Landscaping.' In *Deleuze | Guattari & Ecology*, edited by Bernd Herzogenrath, 251-65. New York: Palgrave Macmillan.

Acker, Paul. 2006. 'Horror and the Maternal in Beowulf.' *PMLA* 121.3: 702-16.

Adams, Carol. 2015. *The Sexual Politics of Meat: A Feminist-Vegetarian Critical Theory*. London: Bloomsbury Academic.

Alaimo, Stacy. 2013. 'Violet-Black.' In *Prismatic Ecology: Ecotheory Beyond Green*, edited by Jeffrey Jerome Cohen, 233-51. Minneapolis: University of Minnesota Press.

Alaimo, Stacy. 2014. 'Feminist Science Studies and Ecocriticism.' In *The Oxford Handbook of Ecocriticism*, edited by Greg Garrard, 188-204. Oxford: Oxford University Press.

Allen, Valerie. 2014. 'Matter.' In *Inhuman Nature*, edited by Jeffrey Jerome Cohen, 61-78. Washington, DC: Oliphaunt Books.

Armbruster, Karla. 1998. '"Buffalo Gals, Won't You Come Out Tonight?": A Call for Boundary-Crossing in Ecofeminist Literary Criticism.' In *Ecofeminist Literary Criticism: Theory, Interpretation, Pedagogy*, edited by Greta Gaard and Patrick D. Murphy, 97-105. Urbana: University of Illinois Press.

Armbruster, Karla, and Kathleen R. Wallace, eds. 2001. *Beyond Nature Writing: Expanding the Boundaries of Ecocriticism*. Charlottesville: University Press of Virginia.

Ashcroft, Bill, Gareth Griffiths, and Helen Tiffin. 2002. *The Empire Writes Back: Theory and Practice in Post-Colonial Literatures*, 2nd edition. New York: Routledge.

Banham, Debby, and Rosamond Faith. 2014. *Anglo-Saxon Farms and Farming*. Oxford: Oxford University Press.

Barndt, Kerstin. 2010. 'Memory Traces of an Abandoned Set of Futures.' In *Ruins of Modernity*, edited by Julia Hell and Andreas Schönle, 270-93. Durham: Duke University Press.

Beasley-Murray, Jon. 2010. 'Vilcashuamán: Telling Stories in Ruins.' In *Ruins of Modernity*, edited by Julia Hell and Andreas Schönle, 213-31. Durham: Duke University Press.

Bennett, Michael, and David W. Teague, eds. 1999. *The Nature of Cities: Ecocriticism and Urban Environments*. Tucson: University of Arizona Press.

Berman, Russell A. 2010. 'Democratic Destruction: Ruins and Emancipation in the American Tradition.' In *Ruins of Modernity*, edited by Julia Hell and Andreas Schönle, 104-17. Durham: Duke University Press.

Bitterli, Dieter. 2009. *Say What I Am Called: The Old English Riddles of the Exeter Book and the Anglo-Latin Riddle Tradition*. Toronto: University of Toronto Press.

Bjork, Robert E., and John D. Niles, eds. 1997. *Beowulf Handbook*. Lincoln: University of Nebraska Press.

Bolintineanu, Alexandra. 2009. 'The Land of the Mermedonians in the Old English *Andreas*'. *Neophilologus* 93: 149-64.

Bosworth, Joseph. 2010. *An Anglo-Saxon Dictionary Online*, edited by Thomas Northcote Toller and others. Compiled by Sean Christ and Ondřej Tichý. Faculty of Arts, Charles University in Prague. Accessed 31 July 2016. http://www.bosworthtoller.com/026643.

Brady, Lindy. 2010. 'Echoes of Britons on a Fenland Frontier in the Old English *Andreas*.' *Review of English Studies* 61: 669-89.

Branch, Michael P. 2002. 'Saving All the Pieces: The Place of Textual Editing in Ecocriticism.' In *The Greening of Literary Scholarship*, edited by Steven Rosendale, 3-25. Iowa City: University of Iowa Press.

Branch, Michael P., and Scott Slovic, eds. 2003. *ISLE Reader: Ecocriticism, 1993-2003* Athens: University of Georgia Press.

Brown, Bill. 2001. 'Thing Theory.' *Critical Inquiry* 28.1: 1-22.

Bryant, Levi. 2013. 'Black.' In *Prismatic Ecologies*, edited by Jeffrey J. Cohen, 290-310. Minneapolis: University of Minnesota Press.

Brown, Carleton. 1938. '*Beowulf* and the *Blickling Homilies* and Some Textual Notes.' *PMLA* 53: 905-15.

Brown, Elsa Barkley. 1989. 'African-American Women's Quilting: A Framework for Conceptualizing and Teaching African-American Women's History.' *Signs* 14.4: 921-29.

Buell, Lawrence. 1996. *The Environmental Imagination: Thoreau, Nature Writing and the Formation of American Culture*. Cambridge: Harvard University Press.

— 2005. *The Future of Environmental Criticism: Environmental Crisis and Literary Imagination*. Malden: Wiley-Blackwell.

Bynum, Caroline Walker. 1991. *Fragmentation and Redemption: Essays on Gender and the Human Body in Medieval Religion*. New York: Urzone Publishers.

Carson, Rachel. 2002 [1962]. *Silent Spring: Fortieth Anniversary Edition*. New York: Houghton Mifflin.

Chance, Jane. 1990. 'The Structural Unity of Beowulf: The Problem of Grendel's Mother.' In *New Readings on Old English Literature*, edited by Helen Damico and Alexandra Hennessey Olsen, 248-61. Bloomington: University of Indiana Press.

Chazelle, Celia, and others. 2012. Introduction to *Why the Middle Ages Matter*, edited by Celia Chazelle and others, 1-14. New York: Routledge.

Claborn, John. 2014. 'W.E.B. Du Bois at the Grand Canyon: Nature, History and Race in *Darkwater*.' In *The Oxford Handbook of Ecocriticism*, edited by Greg Garrard, 118-31. Oxford: Oxford University Press.

Clark, Stephanie. 2012. 'A More Permanent Homeland: Land Tenure in *Guthlac A*.' *Anglo-Saxon England* 40: 75-102.

Clarke, Catherine A.M. 2006. *Literary Landscapes and the Idea of England, 700-1400*. Woodbridge: Boydell and Brewer.

Clark Hall, J.R. 1960. *A Concise Anglo-Saxon Dictionary*, 4th edition. Toronto: University of Toronto Press.

Clunies Ross, Margaret. 1977. 'Textual Territory: The Regional and Genealogical Dynamic of Medieval Icelandic Literary Production.' In *New Medieval Literatures*, edited by Wendy Scase, Rita Copeland, and David Lawton, 9-30. Oxford: Clarendon Press.

— 1998. 'Land-Taking and Text-Making in Medieval Iceland.' In *Text and Territory: Geographical Imagination in the European Middle Ages*, edited by Sylvia Tomasch and Sealy Gilles, 159-84. Philadelphia: University of Pennsylvania Press.

Cohen, Jeffrey Jerome. 2012. Introduction to *Animal, Vegetable, Mineral: Ethics and Objects*, edited by Jeffrey Jerome Cohen, 1-8. Washington, DC: Oliphaunt Books.

— 2013a. Introduction to *Prismatic Ecology: Ecotheory Beyond Green*, edited by Jeffrey Jerome Cohen, xv-xxxv. Minneapolis: University of Minnesota Press.

—, ed. 2013b. *Prismatic Ecology: Ecotheory Beyond Green*. Minneapolis: University of Minneapolis Press.

— 2014. Introduction to *Inhuman Nature*, edited by Jeffrey Jerome Cohen, i-x. Washington, DC: Oliphaunt Books.

— 2015. *Stone: An Ecology of the Inhuman*. Minneapolis: University of Minnesota Press.

Cohen, Jeffrey Jerome and Lowell Duckert, eds. 2015. Introduction to *Elemental Ecocriticism: Thinking with Earth, Air, Water and Fire*, edited by Jeffrey Jerome Cohen and Lowell Duckert, 1-26. Minneapolis: University of Minnesota Press.

Collins, Patricia Hill. 1990. *Black Feminist Thought: Knowledge, Consciousness and the Politics of Empowerment*. New York: Routledge.

Composing Disability: Crip Ecologies (Symposium). George Washington University, 7-8 April 2016. Web. Accessed 1 August 2016. http://composingdisability.blogspot.com.

'Could Offa's Dyke Be Someone Else's?' *British Heritage* 35.4 (2014): 10.

Crane, Kylie. 2012. *Myths of Wilderness in Contemporary Prose Texts: Environmental Postcolonialism in Australia and Canada*. New York: Palgrave Macmillan.

Crane, Susan. 2012. *Animal Encounters: Contacts and Concepts in Medieval Britain*. Philadelphia: University of Pennsylvania Press.

Cronon, William. 1995. 'The Trouble with Wilderness; or, Getting Back to the Wrong Nature.' In *Uncommon Ground: Toward Reinventing Nature*, edited by William Cronon, 69-99. New York: W.W. Norton and Company.

Cullum, P.H., and Katherine Lewis. 2013. *Religious Men and Masculine Identity in the Middle Ages*. Woodbridge: Boydell & Brewer.

Deleuze, Gilles, and Félix Guattari. 2014 [1987]. *A Thousand Plateaus: Capitalism and Schizophrenia*. London: Bloomsbury.

Deloughrey, Elizabeth. 2013. 'Postcolonialism.' In *Oxford Companion to Ecocriticism*, edited by Greg Garrard, 320-40. Oxford: Oxford University Press.

Dendle, Peter. 2014. 'Demons of the Water: Anglo-Saxon Responses to the Gerasene Demoniac.' In *The Maritime World of the Anglo-Saxons*, edited by Stacy S. Klein, William Schipper, and Shannon Lewis-Simpson, 187-207. Tempe: Arizona Center for Medieval And Renaissance Studies.

Derrida, Jacques. 2002. 'The Animal That Therefore I Am (More to Follow).' *Critical Inquiry* 28: 369-418.

Derrida, Jacques. 2008. *The Animal that Therefore I Am*, edited by Marie-Louie Mallet, translated by David Wills. New York: Fordham University Press.

Dictionary of Old English: A to G, edited by Angus Cameron, Ashley Crandell Amos, and Antonette diPaolo Healey. Toronto: University of Toronto Press, 2009. Accessed 2 August 2016. http://doe.utoronto.ca/pages/index.html.

Dictionary of Old English Web Corpus, compiled by Antonette diPaolo Healey with John Price Wilkin and Xin Xiang. Toronto: Dictionary of Old English Project, 2009.

Dockray-Miller, Mary. 2000. *Motherhood and Mothering in Anglo-Saxon England*. New York: St. Martin's Press.

Donoghue, Daniel. 2004. *Old English Literature: A Short Introduction*. New York: Blackwell.

Doubleday, James F. 2011. 'Grendel's Two Halls.' *Notes & Queries* 58. 1: 8-11.

Douglas, Rebecca. 1998. 'Ecocriticism and Middle English Literature.' *Studies in Medievalism* 10: 136-63.

Drimmer, Sonja. 2014. Syllabus for Graduate Seminar: The Lost Art of Medieval England: Absence and Recovery. Accessed 3 February 2015. https://www.academia.edu/9097020/Drimmer_-_Syllabus_Lost_Art_of_Medieval_Britain_2014.

Du Bois, W.E.B. 1920. 'Of Beauty and Death.' In *Darkwater: Voices from Within the Veil*, 104-16. New York; Harcourt Brace.

Ellard, Donna Beth. 2011. 'Going Interspecies, Going Interlingual, and Flying Away with the *Phoenix*.' *Exemplaria* 23.3: 268-92.

Estes, Heide. 2007. 'Abraham and the Northmen in *Genesis A*: Alfredian Translations and Ninth-Century Politics.' *Medievalia et Humanistica* 33: 1-13.

— 2012. 'Anglo-Saxon Biblical Lore: An Edition.' *English Studies* 93.6: 623-51.

Evan, Peter D. 2013. 'Word-Play as Evidence for the Date of *Durham*.' *Medium Ævum* 82.2: 314-17.

Fell, Christine. 1991. 'Perceptions of Transience.' In *The Cambridge Companion to Old English Literature*, edited by Malcolm Godden and Michael Lapidge, 172-98. Cambridge: Cambridge University Press.

Ferhatović, Denis. 2010. '*Burh* and *Beam*, Burning Bright: A Study in the Poetic Imagination of the Old English *Exodus*.' *Neophilologus* 94: 509-22.

Foster, John Bellamy. 2000. *Marx's Ecology: Materialism and Nature*. New York: Monthly Review Press.

Fraiman, Susan. 2012. 'Pussy Panic versus Liking Animals: Tracking Gender in Animal Studies.' *Critical Inquiry* 39: 89-115.

Freud, Sigmund. 2002. *The Joke and its Relationship to the Unconscious*, translated by Joyce Crick. New York: Penguin.

Furman, Andrew. 2003. 'No Trees, Please, We're Jewish.' In *The Isle Reader: Ecocriticism, 1993-2003*, edited by Michael P. Branch and Scott Slovic, 49-71. Athens: University of Georgia Press.

Gaard, Greta. 2010. 'New Directions for Ecofeminism.' *Interdisciplinary Studies in Literature and Environment* 17: 1-23.

Garrard, Greg. 1998. 'Heidegger, Heaney and the Problem of Dwelling.' In *Writing the Environment: Ecocriticism and Literature*, edited by Richard Kerridge and Neil Sammels, 167-81. London: Zed.

— 2012. *Ecocriticism: The New Critical Idiom*. 2nd edition. New York: Routledge.

— 2014. Introduction to *The Oxford Handbook of Ecocriticism*, edited by Greg Garrard, 1-24. Oxford: Oxford University Press.

Gelling, Margaret. 2002. 'The Landscape of *Beowulf*.' *Anglo-Saxon England* 31: 7-11.

Giblett, Rod. 1996. *Postmodern Wetlands: Culture, History, Ecology*. Edinburgh: Edinburgh University Press.

Glotfelty, Cheryll, and Harold Fromm, eds. 1996. *Ecocriticism Reader: Landmarks in Literary Ecology*. Athens: University of Georgia Press.

Godden, M.R. 2002. 'The Sources of Ælfric's Lives of Saints I Nativity of Christ (Cameron C.B.1.3.2).' In *Fontes Anglo-Saxonici: World Wide Web Register*. Accessed July 2011. http://fontes.english.ox.ac.uk.

Goodbody, Axel. 2011. 'Sense of Place and Lieu de Mémoire: A Cultural Memory Approach to Environmental Texts.' In *Ecocritical Theory: New European Approaches*, edited by Axel Goodbody and Kate Rigby, 55-67. Morgantown: University of Virginia Press.

Gneuss, Helmut. 2001. *Handlist of Anglo-Saxon Manuscripts and Manuscript Fragments Written or Owned in England up to 1100*. Tempe: Arizona Center for Medieval and Renaissance Studies.

Grewe-Volpp, Christa. 2006. 'How to Speak the Unspeakable: The Aesthetics of the Voice of Nature.' *Anglia* 124.1: 122-43.

Grossi, Joseph. 2012. 'Preserving the Future in the Old English *Durham*.' *Journal of English and Germanic Philology* 111.1: 42-73.

Haggard, H. Rider. 1907. *King Solomon's Mines*. London: Cassell.

Halttunen, Karen. 2006. 'Groundwork: American Studies in Place.' *American Quarterly* 58.1: 1-15.

Hamer, Richard, ed. and trans. 1970. *A Choice of Anglo-Saxon Verse*. London: Faber & Faber.

Harman, Graham. 2013. 'Gold.' In *Prismatic Ecology: Ecotheory Beyond Green*, edited by Jeffrey Jerome Cohen, 106-23. Minneapolis: University of Minnesota Press.

Harris, Anne F. 2014. 'Hewn.' In *Inhuman Nature*, edited by Jeffrey Jerome Cohen, 17-38. Washington, DC: Oliphaunt Books.

Harris, Jonathan Gil. 2012. 'Animal, Vegetable, Mineral: Twenty Questions.' In *Animal, Vegetable, Mineral: Ethics and Objects*, edited by Jeffrey Jerome Cohen, 285-95. Washington, DC: Oliphaunt Books.

Head, Dominic. 1998. 'The (Im)possibility of Ecocriticism.' In *Writing the Environment: Ecocriticism and Literature*, edited by Richard Kerridge and Neil Sammels, 27-39. London: Zed Books.

Heaney, Seamus. 2000. *Beowulf: A New Verse Translation*. New York: W.W. Norton & Company.

Heise, Ursula K. 1999. 'Forum on Literatures of the Environment.' *PMLA* 114: 1097.

Hell, Julia. 2010. 'Imperial Ruin Gazers, Or Why Did Scipio Weep?' In *Ruins of Modernity*, edited by Julia Hell and Andreas Schönle, 169-92. Durham: Duke University Press.

Hell, Julia, and Andreas Schönle. 2010. Introduction to *Ruins of Modernity*, edited by Julia Hell and Andreas Schönle, 2-14. Durham: Duke University Press.

Herzogenrath, Bernd. 2009. 'Nature | Geophilosopy | Machinics | Ecosophy.' In *Deleuze | Guattari & Ecology*, edited by Bernd Herzogenrath, 1-22. New York: Palgrave Macmillan.

Hiatt, Alfred. 2009. '*Beowulf* Off the Map.' *Anglo-Saxon England* 38: 11-40.

Hills, Catherine M. 1997. 'Beowulf and Archaeology.' In *The Beowulf Handbook*, edited by Robert E. Bjork and John D Niles, 291-310. Lincoln: University of Nebraska.

Hiltner, Ken, ed. 2014. *Ecocriticism: The Essential Reader*. New York: Routledge.

Holsinger, Bruce. 2009. 'Of Pigs and Parchment: Medieval Studies and the Coming of the Animal.' *PMLA* 124.2: 616-23.

Holt, Richard. 2000. 'Society and Population 600-1300.' In *Cambridge Urban History of Britain Volume I: 600-1540*, edited by D.M. Palliser, 79-104. Cambridge: Cambridge University Press.

Honegger, Thomas. 1998. 'Form and Function: The Beasts of Battle Revisited', *English Studies* 79.4: 289-98.

Hooke, Della. 1998. *The Landscape of Anglo-Saxon England*. London: Leicester University Press.

— 2010. *Trees in Anglo-Saxon England: Literature, Lore and Landscape*. Woodbridge: Boydell and Brewer.

Horváth, Ágnes Réffy. 2000. 'Saint Guthlac, the Warrior of God in the Guthlac Poems of the Exeter Book.' *AnaChronisT* 6: 1-28. Accessed 24 March 2015. http://seas3.elte.hu/anachronist/index2.htm.

Howe, Nicholas. 1989. *Migration and Mythmaking in Anglo-Saxon England*. New Haven: University of Notre Dame Press.

— 2002. 'The Landscape of Anglo-Saxon England: Inherited, Invented, Imagined.' In *Inventing Medieval Landscapes: Senses of Place in Western Europe*, edited by John Howe and Michael Wolfe, 91-112. Gainesville: University Press of Florida.

— 2008. *Writing the Map of Anglo-Saxon England: Essays in Cultural Geography*. New Haven: Yale University Press.

Howlett, D.R. 2008. 'Two Old English Encomia.' *English Studies* 57.4: 289-93.

Huggan, Graham, and Helen Tiffin, 2010. *Postcolonial Ecocriticism: Literature, Animals Environment*. New York: Routledge.

Huyssen, Andreas. 2010. 'Authentic Ruins: Products of Modernity.' In *Ruins of Modernity*, edited by Julia Hell and Andreas Schönle, 17-28. Durham: Duke University Press.

Iovino, Serenella, and Serpil Opperman. 2013. 'Onword: After Green Ecologies: Prismatic Visions.' In *Prismatic Ecology: Ecotheory Beyond Green*, edited by Jeffrey Jerome Cohen, 328-36. Minneapolis: University of Minnesota Press.

Irving, Edward B. Jr. 1997. 'Christian and Pagan Elements.' In *A Beowulf Handbook*, edited by Robert E. Bjork and John D. Niles, 175-92. Lincoln: University of Nebraska Press.

Johnson, David F. 2008. 'Spiritual Combat and the Land of Canaan in *Guthlac A*.' In *Intertexts: Studies in Anglo-Saxon Culture Presented to Paul E. Szarmach*, edited by Virginia Blanton and Helene Scheck, 307-17. Tempe, AZ: ACMRS.

Kay, Sarah. 2011. 'Legible Skins: Animals and the Ethics of Medieval Reading.' *Postmedieval* 2.1: 13-32.

Keene, Derek. 'London from the post-Roman Period to 1300.' In *Cambridge Urban History of Britain Volume I: 600-1540*, edited by D.M. Palliser, 187-216. Cambridge: Cambridge University Press.

Ker, N.R. 1977. *Catalogue of Manuscripts Containing Anglo-Saxon*. Oxford: Clarendon Press.

Kerridge, Richard. 2014. 'Ecocritical Approaches to Literary Form and Genre: Urgency, Depth, Provisionality, Temporality.' In *The Oxford Handbook of Ecocriticism*, edited by Greg Garrard, 361-76. Oxford: Oxford University Press.

Kerridge, Richard, and Neil Sammels, eds. 1998. *Writing the Environment: Ecocriticism and Literature*. London: Zed Books.

Kiser, Lisa. 2009. '"Mak's Heirs": Sheep and Humans in the Pastoral Ecology of the Towneley First and Second Shepherds' Plays.' *Journal of English and Germanic Philology* 108.3: 335-59.

Klein, Stacy S. 2006. 'Gender and the Nature of Exile in the Old English Elegies.' In *A Place to Believe In: Locating Medieval Landscapes*, edited by Clare A. Lees and Gillian R. Overing, 113-31. University Park: The Pennsylvania State University Press.

Kohn, Eduardo. 2013. *How Forests Think: Toward an Anthropology Beyond the Human*. Berkeley: University of California Press.

Kordecki, Lesley. 2003. 'Ecofeminism and the Father of English Poetry: Chaucer's *Parliament of Fowls*.' *Interdisciplinary Studies in Literature and Environment* 13.1: 97-114.

Kristeva, Julia. 1982. *Powers of Horror: An Essay on Abjection*, translated by Leon S. Roudiez. New York: Columbia University Press.

Lampert, Lisa. 2004. *Gender and Jewish Difference from Paul to Shakespeare*. Philadelphia: University of Pennsylvania Press.

Lapidge, Michael. 2006. *The Anglo-Saxon Library*. Oxford: Oxford University Press.

Lerer, Seth. 2002. 'Old English and its Afterlife.' In *Cambridge History of Medieval English Literature*, edited by David Wallace, 1-34. Cambridge: Cambridge University Press.

— 2010. '"On Fagne Flor": The Postcolonial *Beowulf*, from Heorot to Heaney.' In *Postcolonial Approaches to the European Middle Ages: Translating Cultures*, edited by Ananya Jahanara Kabir and Deanne Williams, 77-102. Cambridge: Cambridge University Press, 2010.

Liuzza, Roy. 2003. 'The Tower of Babel: *The Wanderer* and the Ruins of History.' *Studies in the Literary Imagination* 36.1: 1-35.

Lousley, Cheryl. 2014. 'Ecocriticism and the Politics of Representation.' In *The Oxford Handbook of Ecocriticism*, edited by Greg Garrard, 155-71. Oxford: Oxford University Press.

Low, Matt. 2009. '"Heard gripe hruson" (The hard grip of the earth), Ecopoetry and the Anglo-Saxon Elegy.' *Mosaic* 4.3: 1-18.

Magoun, Francis Peabody. 1955. 'The Theme of the Beasts of Battle in Anglo-Saxon Poetry.' *Neuphilologische Mitteilungen* 56: 81-90.

Maskit, Jonathan. 2009. 'Subjectivity, Desire, and the Problem of Consumption.' In *Deleuze | Guattari & Ecology*, edited by Bernd Herzogenrath, 129-44. New York: Palgrave Macmillan.

Mellor, Leo. 2014. 'The Lure of Wilderness.' In *Cambridge Companion to Literature and the Environment*, edited by Louise Westling, 104-18. Cambridge: Cambridge University Press.

Menely, Tobias, and Margaret Ronda. 2013. 'Red.' In *Prismatic Ecology: Ecotheory Beyond Green*, edited by Jeffrey Jerome Cohen, 22-41. Minneapolis: University of Minnesota Press.

Middle English Dictionary, edited by Frances McSparran and others. University of Michigan, 2001. https://quod.lib.umich.edu/m/med.

Momma, Haruko. 2005. 'The Education of Beowulf and the Affair of the Leisure Class.' In *Verbal Encounters: Anglo-Saxon and Norse Studies for Roberta Frank*, edited by Antonia Harbus and Russell Poole, 163-82. Toronto: University of Toronto Press.

Monge, Susana Fidalgo. 1999. 'The Sea in *Beowulf*, *The Wanderer*, and *The Seafarer*: On Semantic Fields and Mediterranean Limitations.' *SELIM: Journal of the Spanish Society for Medieval English Language and Literature* 9: 155-62.

Montroso, Alan. 2014. 'Human.' In *Inhuman Nature*, edited by Jeffrey Jerome Cohen, 39-59. Washington, DC: Oliphaunt Books.

Mortimer-Sandilands, Catriona, and Bruce Erickson, eds. 2010. *Queer Ecologies: Sex, Nature, Politics*. Bloomington: Indiana University Press.

Morton, Timothy. 2013a. *Hyperobjects: Philosophy and Ecology after the End of the World*. Minneapolis: University of Minnesota Press.

— 2013b. 'Treating Objects Like Women: Feminist Ontology and the Question of Essence.' In *International Perspectives in Feminist Ecocriticism*, edited by Greta Gaard, Serpil Opperman and Simon Estok, 56-69. New York: Routledge.

Müller, Timo. 2011. 'From Literary Anthropology to Cultural Ecology: German Ecocritical Theory Since Wolfgang Iser.' In *Ecocritical Theory: New European Approaches*, edited by Axel Goodbody and Kate Rigby, 71-83. Morgantown: University of Virginia Press.

Murphy, Patrick D. 1991. 'Ground, Pivot, Motion: Ecofeminist Theory, Dialogics, and Literary Practice.' *Hypatia* 6.1: 146-61.

— 1995. *Literature, Nature, and Other: Ecofeminist Critiques*. Albany: State University of New York Press.

— 1999. 'Forum on Literatures of the Environment.' *PMLA* 114: 1099.

Murphy, Patrick J. 2011. *Unriddling the Exeter Riddles*. State College: Penn State University Press.

Neville, Jennifer. 1999. *Representations of the Natural World in Old English Poetry*. Cambridge: Cambridge University Press.

Niles, John D. 1997. 'Myth and History.' In *The Beowulf Handbook*, edited by Robert E. Bjork and John D. Niles, 213-32. Lincoln: University of Nebraska Press.

— 2006. *Old English Enigmatic Poems and the Play of the Texts*. Turnhout: Brepols.

O'Donnell, Thomas W. 2014. 'The Old English *Durham*, the *Historia de sancto Cuthberto*, and the Unreformed in Late Anglo-Saxon Literature.' *Journal of English and Germanic Philology* 113.2: 131-55.

O'Brien O'Keeffe, Katherine. 1981. '*Beowulf*, lines 702b-836: Transformations and the Limits of the Human.' *Texas Studies in Literature and Language* 23.4: 484-94.

— 2001. 'Guthlac's Crossings.' *Quaestio* 2: 1-26.

Olsen, Karin. 1998. 'The Dichotomy of Land and Sea in The Old English Andreas.' *English Studies* 79.5: 385-94.

Oppermann, Serpil. 2006. 'Theorizing Ecocriticism: Toward a Postmodern Ecocritical Practice.' *Interdisciplinary Studies in Literature and Environment* 13.2: 103-28.

Orchard, Andy. 1985. *Pride and Prodigies: Studies in the Monsters of the Beowulf-Manuscript*. Toronto: University of Toronto Press.

Ortenburg, Veronica. 1990. 'Archbishop Sigeric's Journey to Rome in 990.' *Anglo-Saxon England* 19: 197-246.

Osborne, Marijane. 1992. '"Verbal Sea Charts" and Beowulf's Approach to Denmark.' In *De Gustibus: Essays for Alain Renoir*, edited by John Miles Foley, Chris Womack, and Whitney A. Womack, 441-45. New York: Garland.

Overing, Gillian R., and Marijane Osborne. 1994. *Landscape of Desire: Partial Stories of the Medieval Scandinavian World*. Minneapolis: University of Minnesota Press.

Owen-Crocker, Gale R. 1986. *Dress in Anglo-Saxon England*. Woodbridge: Boydell and Brewer.

Palliser, D.M. 2000. 'The Origins of British Towns.' In *Cambridge Urban History of Britain Volume I: 600-1540*, edited by D.M. Palliser, 18-20. Cambridge: Cambridge University Press.

Parkes, James. 1961. *Conflict of the Church and the Synagogue: A Study of the Origins of Anti-Semitism*. New York: Meridian.

Plumwood, Val. 1993. *Feminism and the Mastery of Nature*. New York: Routledge.

— 2002. *Environmental Culture: The Ecological Crisis of Reason*. New York: Routledge.

Puff, Helmut. 2010. 'Ruins as Models: Displaying Destruction in Postwar Germany.' In *Ruins of Modernity*, edited by Julia Hell and Andreas Schönle, 254-69. Durham: Duke University Press.

Quine, W.V., and J.S. Ullian. 1970. *The Web of Belief*. New York: Random House.

Rackham, Oliver. 1994. 'Trees and Woodland in Anglo-Saxon England: The Documentary Evidence.' In *Environment and Economy in Anglo-Saxon England*, edited by James Rackham, 7-11. York: Council for British Archaeology.

Reichardt, Paul F. 1974. '*Guthlac A* and the Landscape of Spiritual Perfection.' *Neophilologus* 58: 331-38.

Roberts, Jane. 1986. 'The Old English Prose Translation of Felix's *Vita Sancti Guthlaci*.' In *Studies in Earlier Old English Prose*, edited by Paul E. Szarmach, 363-79. Albany: State University of New York Press.

— 1988. '*Guthlac A*: Sources and Source Hunting.' In *Medieval English Studies Presented to George Kane*, edited by Edward Donald Kennedy, Ronald Waldron, and Joseph S. Wittig, 1-18. Woodbridge: Brewer.

Robertson, Kellie. 2012. 'Exemplary Rocks.' In *Animal, Vegetable, Mineral: Ethics and Objects*, edited by Jeffrey Jerome Cohen, 91-122. Washington, DC: Oliphaunt Books.

Rossi-Reder, Andrea. 2004. 'Embodying Christ, Embodying Nation: Ælfric's Accounts of Saints Agatha and Lucy.' In *Sex and Sexuality in Anglo-Saxon England: Essays in Memory of Daniel Gillmore Calder*, edited by Carol Pasternack and Lisa M.C. Weston, 183-202. Tempe: Arizona Center for Medieval and Renaissance Studies.

Rudd, Gillian. 2007. *Greenery: Ecocritical Readings of Late Medieval English Literature*. Manchester: Manchester University Press.

— 2014. 'Being Green in Late Medieval English Literature.' In *The Oxford Handbook of Ecocriticism*, edited by Greg Garrard, 27-39. Oxford: Oxford University Press.

Ruether, Rosemary Radford. 1974. *Faith and Fratricide: The Theological Roots of Anti-Semitism*. Minneapolis: Seabury Press.

Rulon-Miller, Nina. 2000. 'Sexual Humor and Fettered Desire in Exeter Book Riddle 12.' In *Humor in Anglo-Saxon Literature*, edited by Jonathan Wilcox, 99-126. Woodbridge: D.S. Brewer.

Salih, Sara. 2014. 'Vegans on the Verge of a Nervous Breakdown.' In *Rise of Critical Animal Studies: From the Margins to the Centre*, edited by Nik Taylor and Richard Twine, 52-68. New York: Routledge.

Salvador-Bello, Mercedes. 2011. 'The Sexual Riddle Type in Aldhelm's *Enigmata*, the *Exeter Book*, and Early Medieval Latin.' *Philological Quarterly* 90.4: 357-85.

Schücking, Levin L. 1915. *Untersuchungen zur Bedeutungslehre der angelsächsischen Dichtersprache*. Heidelberg: Carl Winters Universitätsbuchhandlung.

Sharma, Manish. 2002. 'A Reconsideration of the Structure of *Guthlac A*: The Extremes of Saintliness'. *Journal of English and Germanic Philology* 101.2: 185-200.

Shook, Laurence K. 1960. 'The Burial Mound in *Guthlac A*' *Modern Philology* 58.1: 1-10.

Siewers, Alfred K. 2003. 'Landscapes of Conversion: Guthlac's Mound and Grendel's Mere as Expressions of Anglo-Saxon Nation-Building.' *Viator: Medieval and Renaissance Studies* 34.7: 1-39.

— 2009. *Strange Beauty: Ecocritical Approaches to Early Medieval Landscape*. New York: Palgrave Macmillan.

— 2012. 'Desert Islands: Europe's Atlantic Archipelago as Ascetic Landscape.' In *Studies in the Medieval Atlantic*, edited by Benjamin Hudson, 35-63. New York: Palgrave Macmillan.

— 2014. 'Trees.' In *Inhuman Nature*, edited by Jeffrey Jerome Cohen, 101-13. Washington, DC: Oliphaunt Books.

Singer, Peter. 2015. *Animal Liberation: The Definitive Classic of the Animal Movement*. 40th Anniversary Edition. New York: Open Road Media.

Slicer, Deborah. 1998. 'Toward an Ecofeminist Standpoint Theory: Bodies as Grounds.' In *Ecofeminist Literary Criticism: Theory, Interpretation, Pedagogy*, edited by Greta Gaard and Patrick D. Murphy, 49-73. Urbana: University of Illinois Press.

Slocombe, Will. 2005. 'Littered with Meaning: The Problem of Sign Pollution in Postmodern, Post-Structuralist and Ecocritical Thought.' *Textual Practice* 19.4: 493-508.

Slovic, Scott. 2008. *Going Away to Think: Engagement, Retreat, and Ecocritical Responsibility.* Reno: University of Nevada Press.

Smith, D.K. 2000. 'Humor in Hiding: Laughter Between the Sheets in the *Exeter Book Riddles*.' In *Humor in Anglo-Saxon Literature*, edited by Jonathan Wilcox, 79-98. Woodbridge: D.S. Brewer.

Smith, Scott T. 2012. *Land and Book: Literature and Land Tenure in Anglo-Saxon England*. Toronto: University of Toronto Press.

Stanbury, Sarah. 2004. 'EcoChaucer: Green Ethics and Medieval Nature.' *Chaucer Review* 39.1: 1-16.

Stark, Judith Chelius. 2007. Introduction to *Feminist Interpretations of Augustine*, edited by Judith Chelius Stark, 1-44. University Park: Pennsylvania University Press.

Steel, Karl. 2008. 'How to Make A Human.' *Exemplaria* 20.1: 3-27.

— 2011. *How to Make a Human: Animals and Violence in the Middle Ages*. Columbus: Ohio State University Press.

Stewart, Anne Harleman. 1983. 'Double Entendre in the Old English *Riddles*.' *Lore and Language* 3.8: 39-52.

Stockton, Will. 2013. 'Beige.' In *Prismatic Ecology: Ecotheory Beyond Green*, edited by Jeffrey Jerome Cohen, 170-92. Minneapolis: University of Minnesota Press.

Taylor, Nik, and Richard Twine, eds. 2014. *Rise of Critical Animal Studies: From the Margins to the Centre*. New York: Routledge.

Thesaurus of Old English, edited by Jane Roberts, Christian Kay, and Lynne Grundy. University of Glasgow, 2005. Accessed 1 August 2016. http://oldenglishthesaurus.arts.gla.ac.uk.

Trautmann, Moritz. 1915. *Die Rätsel des Exeterbuchs: Die altenglischen Rätsel*. Heidelberg: Winter.

Treharne, Elaine. 2013. 'Fleshing out the Text: The Transcendent Manuscript in the Digital Age.' *Postmedieval* 4: 465-78.

Treharne, Elaine and Mary Swan, eds. 2000. *Rewriting Old English in the Twelfth Century*. Cambridge: Cambridge University Press.

Tuan, Yi-Fu. 2001. *Space and Place: The Perspective of Experience*. Minneapolis: University of Minnesota Press.

Tupper, Frederick Jr. 1903. 'Originals and Analogues of the Exeter Book Riddles.' *Modern Language Notes* 18.4: 97-106.

Watson, Robert. 2014. 'Shadows of the Renaissance.' In *The Oxford Handbook of Ecocriticism*, edited by Greg Garrard, 40-59. Oxford: Oxford University Press.

Wentersdorf, Karl P. 1978. '*Guthlac A*: The Battle for the Beorg.' *Neophilologus* 62: 135-42.

— 2007. 'The *Beowulf*-Poet's Vision of Heorot.' *Studies in Philology* 104.4: 409-26.

Westling, Louise, ed. 2014. *Cambridge Companion to Literature and the Environment*. Cambridge: Cambridge University Press.

White, Lynn Jr., 1996. 'The Historical Roots of our Ecological Crisis.' In *Ecocriticism Reader: Landmarks in Literary Ecology*, edited by Cheryll Glotfelty and Harold Fromm, 3-14. Athens: University of Georgia Press.

Wickham-Crowley, Kelley M. 2006. 'Living on the Ecg: The Mutable Boundaries of Land and Water in Anglo-Saxon Contexts.' In *A Place to Believe In: Locating Medieval Landscapes*, edited by Clare A. Lees and Gillian R. Overing, 85-100. University Park: The Pennsylvania State University Press.

Williamson, Tom. 2003. *Shaping Medieval Landscapes: Settlement, Society, Environment*. Macclesfield: Windgather Press.

Wilcox, Jonathan. 2012. 'Digging for New Meanings: Uncovering a Postcolonial *Beowulf*.' In *Language Studies: Stretching the Boundaries*, edited by Andrew Littlejohn and Sandhya Rao Mehta, 147-61. Newcastle upon Tyne: Cambridge Scholars Publishing.

Wright, Laura. 2010. *'Wilderness into Civilized Shapes': Reading the Postcolonial Environment*. Athens: University of Georgia Press.

Yamamoto, Dorothy. 2000. *The Boundaries of the Human in Medieval English Literature* Oxford: Oxford University Press.

Zapf, Hubert. 2006. 'The State of Ecocriticism and the Function of Literature as Cultural Ecology.' In *Nature in Literary and Cultural Studies: Transatlantic Conversations on Ecocriticism*, edited by Catrin Gersdorf and Sylvia Mayer, 49-70. New York: Rodopi.

Ziegelmaier, Gregory. 1969. 'God and Nature in the Beowulf Poem.' *American Benedictine Review* 20: 250-58.

Index

For Product Safety Concerns and Information please contact our
EU representative GPSR@taylorandfrancis.com Taylor & Francis
Verlag GmbH, Kaufingerstraße 24, 80331 München, Germany